Sediment Transport in the Lower Snake and Clearwater River Basins, Idaho and Washington, 2008–11

By Gregory M. Clark, Ryan L. Fosness, and Molly S. Wood

Prepared in cooperation with the U.S. Army Corps of Engineers

Scientific Investigation Report 2013–5083

U.S. Department of the Interior
U.S. Geological Survey

U.S. Department of the Interior
SALLY JEWELL, Secretary

U.S. Geological Survey
Suzette M. Kimball, Acting Director

U.S. Geological Survey, Reston, Virginia: 2013

For more information on the USGS—the Federal source for science about the Earth, its natural and living resources, natural hazards, and the environment, visit http://www.usgs.gov or call 1–888–ASK–USGS.

For an overview of USGS information products, including maps, imagery, and publications, visit http://www.usgs.gov/pubprod

To order this and other USGS information products, visit http://store.usgs.gov

Suggested citation:
Clark, G.M., Fosness, R.L., and Wood, M.S., 2013, Sediment transport in the lower Snake and Clearwater River Basins, Idaho and Washington, 2008–11: U.S. Geological Survey Scientific Investigations Report 2013-5083, 56 p.

Contents

Abstract...1

Introduction...2

    Purpose and Scope 5

Description of Study Area 5

    Salmon River Basin..5

    Snake River Downstream of Hells Canyon Dam..5

    Grande Ronde River Basin..6

    Clearwater River Basin...6

    Palouse River Basin...7

Methods of Data Collection and Analysis 7

    Streamflow..8

    Suspended Sediment and Bedload 8

        Suspended Sediment 8

        Suspended-Sediment Surrogates..11

        Bedload..13

Sediment Transport in the Lower Snake and Clearwater River Basins........................14

    Suspended-Sediment Transport..16

        Suspended Sediment Delivery to Lower Granite Reservoir 28

        Comparison with 1972–79 Study..28

    Suspended-Sediment Surrogates...33

    Bedload...41

Summary...52

Acknowledgments...53

References Cited...54

Figures

1. Map showing lower Snake, Salmon, and Clearwater River Basins, Idaho, Oregon, and Washington..3

2. Map showing land use and location of sediment-sampling stations in the lower Snake, Salmon, and Clearwater River Basins, Idaho, Oregon, and Washington, 2008–11..4

3. Hydrographs showing streamflows at which suspended-sediment and bedload samples were collected from the Snake River near Anatone, Washington, and Clearwater River at Spalding, Idaho, water years 2008–11......................................10

4. Hydrographs showing daily mean streamflow in the Snake River near Anatone, Washington and the Clearwater River at Spalding, Idaho, 2008–11, compared with the 30-year mean for water years 1981–2010...15

5. Suspended-sediment transport curves and 95-percent prediction intervals representing best-fit regression equations for a log-log power fit of total suspended sediment with streamflow in the Salmon River at Whitebird, Idaho, Snake River near Anatone, Washington, South Fork Clearwater River at Stites, Idaho, and Clearwater River at Spalding, Idaho..17

6. Suspended-sediment transport curves representing best-fit regression equations for total suspended-sediment concentrations and total suspended-sediment loads at sampling stations in the lower Snake and Clearwater River Basins, Washington and Idaho...18

7. Boxplots showing estimated total suspended-sediment loads and 95-percent confidence intervals for stations in the lower Snake and Clearwater River Basins, Idaho and Washington, water years 2009–11...24

8. Graph showing mean annual yield of suspended sand and fines in the lower Snake and Clearwater River Basins, Washington and Idaho, water years 2009–11 25

9. Flow charts showing estimated loads of total suspended sediment, total suspended sand, and total suspended fines transported in the lower Snake and Clearwater River Basins, water years 2009–11 26

10. Graph showing suspended-sediment loads delivered monthly to Lower Granite Reservoir from the Snake and Clearwater Rivers, Washington and Idaho, March 2008–September 2011 29

11. Suspended-sediment transport curves for concentrations and loads in the Snake River near Anatone, Washington, and Clearwater River at Spalding, Idaho, for data collected during water years 1972–79 and 2008–11 30

12. Suspended-sediment transport curves for concentrations and loads for suspended fines and suspended sands, Snake River near Anatone, Washington, water years 1972–79 and 2008–11..31

13. Graph showing estimated annual suspended sand and fine loads, Snake River near Anatone, Washington, and Clearwater River at Spalding, Idaho, water years 1972–79 and 2009–11 32

14. Best-fit surrogate models for the Snake River near Anatone, Washington, using the 1.5MHz acoustic Doppler velocity meter and the Clearwater River at Spalding, Idaho, using the 3MHz ADVM, water years 2010–11 34

15. Graphs showing comparison of total monthly sediment load estimated using acoustic backscatter and LOADEST models in the Snake River near Anatone, Washington, and Clearwater River at Spalding, Idaho, water years 2009–11 36

16. Graph showing comparison of suspended-sediment concentrations estimated using acoustic backscatter and a LOADEST model, Snake River near Anatone, Washington, water year 2011 37

Figures

17. Graphs showing comparison of total suspended sediment, suspended sand, and suspended fine concentrations estimated using acoustic backscatter and a LOADEST model during spring snowmelt runoff at Snake River near Anatone, Washington, June 2010 39

18. Graph showing comparison of total suspended-sediment concentrations estimated using acoustic backscatter and a LOADEST model during a hydrologic event at Clearwater River at Spalding, Idaho, April and May 2010 41

19. Flow chart showing estimated bedload transported in the lower Snake and Clearwater River Basins, Washington and Idaho, water years 2009–11 45

20. Bedload transport curves representing best-fit regression equations for selected stations in the lower Snake and Clearwater River Basins, Washington and Idaho 46

21. Bedload transport curves based on samples collected from the Lochsa and Selway Rivers near Lowell, Idaho, during water years 2009–11, and during 1994–97 49

22. Bedload transport curves comparing data collected at Snake River near Anatone, Washington, during water years 1972–79 and 2008–11 51

23. Bedload transport curves comparing data collected at Clearwater River at Spalding, Idaho, water years 1972–79 and 2008–11 52

Tables

1. U.S. Geological Survey streamgaging stations where sediment samples were collected, lower Snake and Clearwater River Basins, Washington and Idaho, water years 2008–11 7

2. Ranges of streamflow sampled and suspended-sediment concentrations in samples collected from the lower Snake and Clearwater River Basins, Washington and Idaho, water years 2008–11 9

3. Regression coefficients and coefficients of determination for models used to estimate concentrations and loads of total suspended sediment and suspended-sediment size fractions at stations in the lower Snake and Clearwater River Basins, Washington and Idaho 12

4. Estimated annual, total, and mean loads of suspended sediment, suspended sand, and suspended fines for stations in the lower Snake and Clearwater River Basins, Washington and Idaho, water years 2009–11 19

5. Summary of acoustic surrogate and LOADEST models used to evaluate suspended-sediment concentrations at the Snake River near Anatone, Washington, and Clearwater River at Spalding, Idaho. 33

6. Comparison of suspended-sediment loads estimated using acoustic backscatter and LOADEST models in the Snake River near Anatone, Washington, and Clearwater River at Spalding, Idaho 35

7. Comparison of suspended-sediment loads during selected hydrologic events calculated using acoustic backscatter and a LOADEST model in the Snake River near Anatone, Washington, and Clearwater River at Spalding, Idaho 38

8. Bedload data from sampling stations in the lower Snake and Clearwater River Basins, Washington and Idaho, water years 2008–11 42

9. Best-fit regression equations for bedload from sampling stations in the lower Snake and Clearwater River Basins, Washington and Idaho, water years 2008–11 46

10. Particle-size distribution for bedload samples collected from stations in the lower Snake and Clearwater River Basins, Washington and Idaho, water years 2008–11 47

Conversion Factors and Datums

Conversion Factors

Inch/Pound to SI

Multiply	By	To obtain
Length		
inch (in.)	25.4	millimeter (mm)
foot (ft)	0.3048	meter (m)
mile (mi)	1.609	kilometer (km)
Area		
square mile (mi^2)	2.590	square kilometer (km^2)
Flow rate		
cubic foot per second (ft^3/s)	0.02832	cubic meter per second (m^3/s)
Mass		
ton, short (2,000 lb)	0.9072	megagram (Mg)
ton per square mile per year [(ton/mi^2)/yr]	0.3502	megagram per square kilometer year [(Mg/km^2)/yr]
ton per day (ton/d)	0.9072	metric ton per day
ton per month (ton/m)	0.9072	metric ton per month
ton per year (ton/yr)	0.9072	metric ton per year

SI to Inch/Pound

Multiply	By	To obtain
Length		
millimeter (mm)	0.03937	inch (in.)
Volume		
liter (L)	0.2642	gallon (gal)
Mass		
gram (g)	0.03527	ounce, avoirdupois (oz)
kilogram per day (kg/d)	2.205	pound per day (lb/d)

Concentrations of chemical constituents in water are given either in milligrams per liter (mg/L) or micrograms per liter (µg/L).

Datums

Vertical coordinate information is referenced to the North American Vertical Datum of 1988 (NADV 88).

Horizontal coordinate information is referenced to North American Datum of 1983 (NAD83).

Altitude, as used in this report, refers to distance above the vertical datum.

Sediment Transport in the Lower Snake and Clearwater River Basins, Idaho and Washington, 2008-11

By Gregory M. Clark, Ryan L. Fosness, and Molly S. Wood

Abstract

Sedimentation is an ongoing maintenance problem for reservoirs, limiting reservoir storage capacity and navigation. Because Lower Granite Reservoir in Washington is the most upstream of the four U.S. Army Corps of Engineers reservoirs on the lower Snake River, it receives and retains the largest amount of sediment. In 2008, in cooperation with the U.S. Army Corps of Engineers, the U.S. Geological Survey began a study to quantify sediment transport to Lower Granite Reservoir. Samples of suspended sediment and bedload were collected from streamgaging stations on the Snake River near Anatone, Washington, and the Clearwater River at Spalding, Idaho. Both streamgages were equipped with an acoustic Doppler velocity meter to evaluate the efficacy of acoustic backscatter for estimating suspended-sediment concentrations and transport. In 2009, sediment sampling was extended to 10 additional locations in tributary watersheds to help identify the dominant source areas for sediment delivery to Lower Granite Reservoir. Suspended-sediment samples were collected 9–15 times per year at each location to encompass a range of streamflow conditions and to capture significant hydrologic events such as peak snowmelt runoff and rain-on-snow. Bedload samples were collected at a subset of stations where the stream conditions were conducive for sampling, and when streamflow was sufficiently high for bedload transport.

At most sampling locations, the concentration of suspended sediment varied by 3–5 orders of magnitude with concentrations directly correlated to streamflow. The largest median concentrations of suspended sediment (100 and 94 mg/L) were in samples collected from stations on the Palouse River at Hooper, Washington, and the Salmon River at White Bird, Idaho, respectively. The smallest median concentrations were in samples collected from the Selway River near Lowell, Idaho (11 mg/L), the Lochsa River near Lowell, Idaho (11 mg/L), the Clearwater River at Orofino, Idaho (13 mg/L), and the Middle Fork Clearwater River at Kooskia, Idaho (15 mg/L). The largest measured concentrations of suspended sediment (3,300 and 1,400 mg/L) during a rain-on-snow event in January 2011 were from samples collected at the Potlatch River near Spalding, Idaho, and the Palouse River at Hooper, Washington, respectively. Generally, samples collected from agricultural watersheds had a high percentage of silt and clay-sized suspended sediment, whereas samples collected from forested watersheds had a high percentage of sand.

During water years 2009–11, Lower Granite Reservoir received about 10 million tons of suspended sediment from the combined loads of the Snake and Clearwater Rivers. The Snake River accounted for about 2.97 million tons per year (about 89 percent) of the total suspended sediment, 1.48 million tons per year (about 90 percent) of the suspended sand, and about 1.52 million tons per year (87 percent) of the suspended silt and clay. Of the suspended sediment transported to Lower Granite Reservoir, the Salmon River accounted for about 51 percent of the total suspended sediment, about 56 percent of the suspended sand, and about 44 percent of the suspended silt and clay. About 6.2 million tons (62 percent) of the sediment contributed to Lower Granite Reservoir during 2009–11 entered during water year 2011, which was characterized by an above average winter snowpack and sustained spring runoff.

A comparison of historical data collected from the Snake River near Anatone with data collected during this study indicates that concentrations of total suspended sediment and suspended sand in the Snake River were significantly smaller during water years 1972–79 than during 2008–11. Most of the increased sediment content in the Snake River is attributable to an increase of sand-size material. During 1972–79, sand accounted for an average of 28 percent of the suspended-sediment load; during 2008–11, sand accounted for an average of 48 percent. Historical data from the Clearwater River at Spalding indicates that the concentrations of total suspended sediment collected during 1972–79 were not significantly different from the concentrations measured during this study. However, the suspended-sand concentrations in the Clearwater River were significantly smaller during 1972–79 than during 2008–11. The increase in suspended-sand concentrations in the Snake and Clearwater Rivers are probably attributable to numerous severe forest fires that burned large areas of central Idaho from 1980–2010.

Acoustic backscatter from an acoustic Doppler velocity meter proved to be an effective method of estimating suspended-sediment concentration and load for most streamflow conditions in the Snake and Clearwater Rivers. Models based on acoustic backscatter were able to simulate most of the variability in suspended-sediment concentrations in the Clearwater River at Spalding (coefficient of determination [R^2]=0.93) and the Snake River near Anatone (R^2=0.92). Acoustic backscatter seems to be especially effective for estimating suspended-sediment concentration and load over short (monthly and single storm event) and long (annual) time scales when sediment load is highly variable. However, during high streamflow events acoustic surrogate tools may be unable to capture the contribution of suspended sand moving near the bottom of the water column and thus, underestimate the total load of suspended sediment.

At the stations where bedload was collected, the particle-size distribution at low streamflows typically was unimodal with sand comprising the dominant particle size. At higher streamflows and during peak bedload discharge, the particle size typically was bimodal and was comprised primarily of sand and coarse gravel. About 55,000 tons of bedload was discharged from the Snake River to Lower Granite Reservoir during water years 2009–11, about 0.62 percent of the total sediment load delivered by the Snake River. About 9,500 tons of bedload was discharged from the Clearwater River to Lower Granite Reservoir during 2009–11, about 0.83 percent of the total sediment load discharged by the Clearwater River during 2009–11.

Introduction

Since construction of the first dam on the lower Snake River, the U.S. Army Corps of Engineers (USACE) has recognized that managing sediment is an ongoing maintenance issue in the reservoirs on the lower Snake River. Historically, the USACE has used dredging to manage accumulated sediment and to maintain a sufficient navigational channel. Recently, however, the USACE determined that managing sediment in the upstream watersheds might be a more effective approach to reducing sediment accumulation in the reservoirs. Although the USACE does not have the authority to manage land outside of the reservoir project boundaries, they can identify and evaluate management strategies that could be implemented on non-USACE property to reduce sediment mobilization and transport (Tetra Tech, 2006).

Because Lower Granite Reservoir is the farthest upstream reservoir on the lower Snake River, it receives and retains the largest amount of sediment. Based on extensive sediment range surveys conducted by the USACE, Lower Granite Reservoir contains an estimated 75 million cubic yards of sediment, with average annual sediment inputs of about

2.3 million cubic yards since impoundment in 1975 (Teasdale, 2010). Because of the accumulated sediment, water depths in the navigation channel near the Ports of Lewiston, Idaho, and Clarkston, Washington, at times are less than the 14-ft authorized depth and in some places are as shallow as 8 ft. Because of this reduction in the depths of the channel and the port berthing areas, some port facilities have been forced to operate at reduced capacity. Sedimentation has caused increased safety risks for the shipping industry; increased risks of grounding and damage to equipment; and decreased efficiencies due to modified approach, loading, and unloading procedures (U.S. Army Corps of Engineers, 2005).

In addition to these navigation issues, deposition of sediment and reduced channel capacity at the confluence of the Snake and Clearwater Rivers have reduced the effectiveness of the levee system that protects the cities of Lewiston and Clarkston from flooding. If allowed to continue, sedimentation may reduce the flow capacity to a point that the Standard Project Flood, an estimated or hypothetical flood that may be expected from the most severe combination of weather and flow conditions that are considered reasonably characteristic of the geographical area, may overtop the levees in Lewiston (U.S. Army Corps of Engineers, 2002).

The USACE asked the U.S. Geological Survey (USGS) to measure sediment transport in the Snake and Clearwater Rivers during construction and following the completion of Lower Granite Dam in 1972. The USGS collected data from the Snake River near Anatone, Washington, and the Clearwater River at Spalding, Idaho, from 1972 through 1979. From these data, the USGS developed sediment rating curves to estimate potential sediment transport and deposition to Lower Granite Reservoir (Jones and Seitz, 1980). Suspended silt and sand dominated the sediment load entering the reservoir, whereas bedload accounted for only about 5 percent. Based on the calculations by Jones and Seitz (1980), the Snake River delivered, on average, about 1.8 million tons per year (80 percent) of the total sediment load entering Lower Granite Reservoir; the Clearwater River accounted for about 0.47 million tons per year (about 20 percent).

In 2005, the USACE published a Notice of Intent (U.S. Army Corps of Engineers, 2005) stating plans to prepare an Environmental Impact Statement for a Programmatic Sediment Management Plan (PSMP) to address sediment management within the four lower Snake River reservoirs. The intent of the PSMP was to identify ways to reduce the amount of sediment entering the reservoirs, how to manage the sediment once it enters the reservoirs, and possible changes to structures or operations that could reduce maintenance yet maintain navigational access to the ports of Lewiston and Clarkston. By using various sediment management measures, the PSMP will become a comprehensive, watershed-level framework for managing sediment movement and deposition while maintaining current uses such as commercial navigation, irrigation withdrawals, recreation, and flow conveyance.

In March 2008, as part of its PSMP, the USACE asked the USGS to start a second sediment-sampling program in the lower Snake and Clearwater River Basins (fig. 1) and to evaluate sediment depositional characteristics in Lower Granite Reservoir. In addition to conventional sampling for suspended sediment and bedload, the USGS equipped streamgages on the Snake River near Anatone, Washington (USGS streamgage 13334300), and the Clearwater River at Spalding, Idaho (streamgage 13342500) (fig. 2) with acoustic Doppler velocity meters (ADVMs) and other instruments to evaluate whether surrogate technologies for measuring suspended-sediment concentration could adequately estimate suspended-sediment transport to Lower Granite Reservoir. In addition to the data described in this report, the USACE also contracted the USGS to evaluate sediment deposition within Lower Granite Reservoir (Williams and others, 2012; Braun and others, 2012).

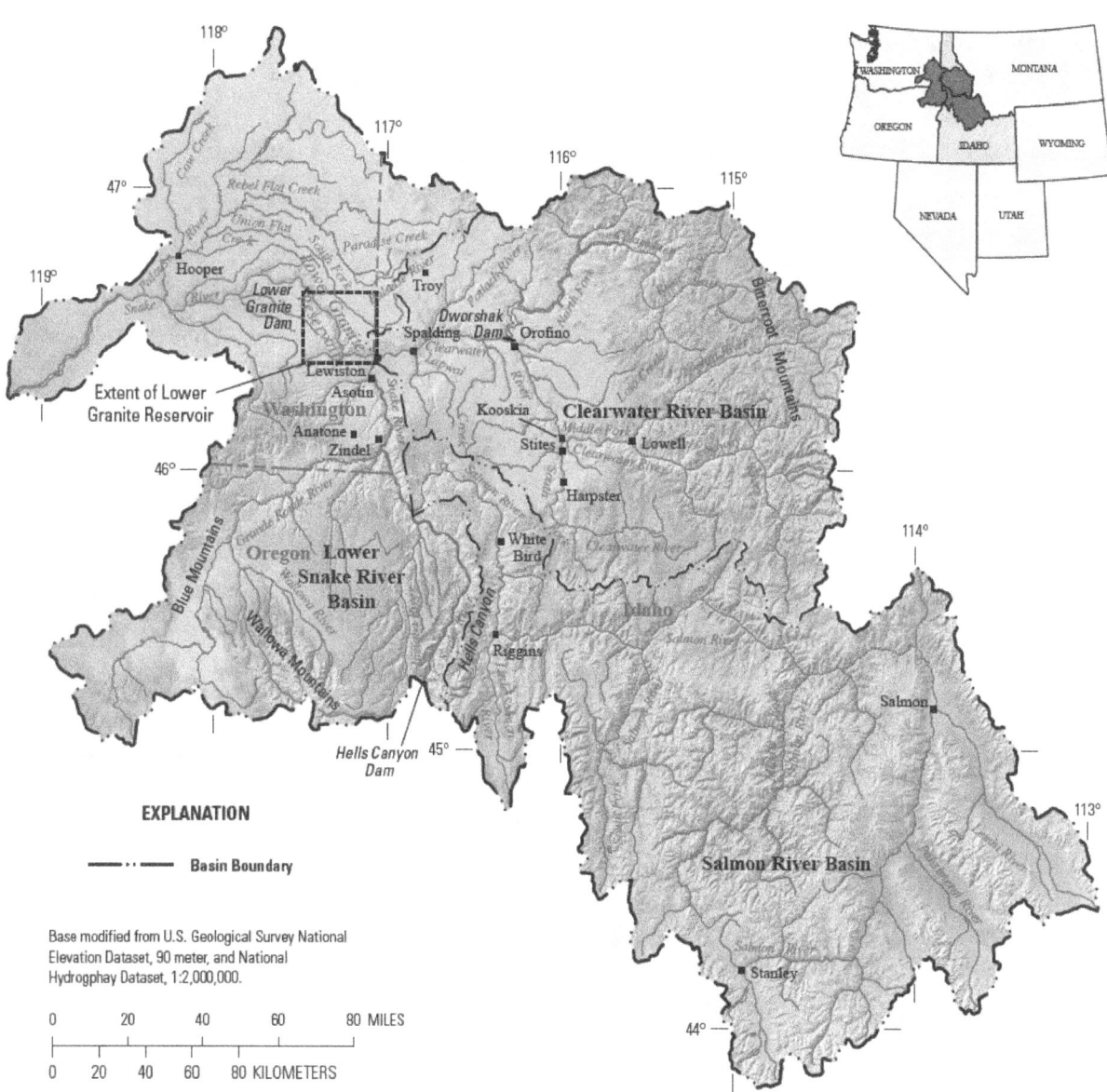

Figure 1.   Lower Snake, Salmon, and Clearwater River Basins, Idaho, Oregon, and Washington.

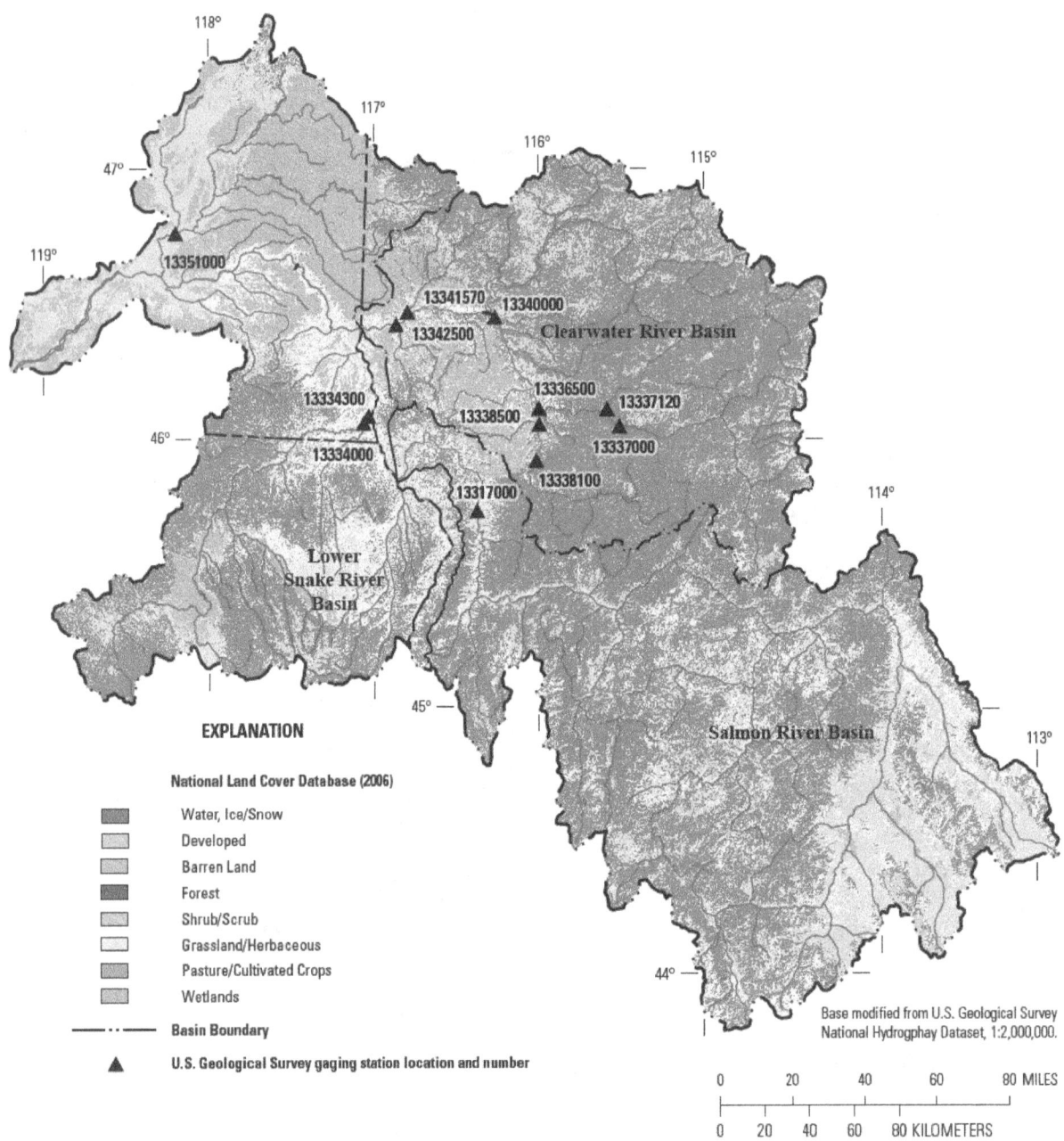

Figure 2.   Land use and location of sediment-sampling stations in the lower Snake, Salmon, and Clearwater River Basins, Idaho, Oregon, and Washington, 2008–11.

Purpose and Scope

This report documents findings based on sediment data collected by the USGS in the lower Snake and Clearwater River Basins from 2008 through 2011. The purpose of this report is to provide information that improves the scientific understanding of the processes affecting sediment generation in the lower Snake and Clearwater River Basins and river hydrodynamics controlling sediment transport, deposition, and retention in Lower Granite Reservoir and downstream in the Snake River. The report provides suspended sediment and bedload data to (1) identify the dominant subbasins contributing sediment to Lower Granite Reservoir, (2) evaluate sediment transport in the Clearwater and lower Snake River Basins, and (3) quantify transport of sediment to Lower Granite Reservoir. Additionally, data collected using acoustic Doppler velocity meters (ADVMs) is summarized and compared with conventionally acquired sediment data to evaluate the efficacy of estimating suspended-sediment concentration (SSC) continuously by means of acoustic surrogates.

Description of Study Area

The drainage basin contributing sediment to Lower Granite Reservoir comprises an area of about 27,000 mi^2 that includes the Snake River Basin from Hells Canyon Dam downstream to the confluence with the Clearwater River and the Clearwater River Basin (fig. 1). The three dams constituting the Hells Canyon Complex (completed in 1967) on the Snake River effectively trap sediment transported from areas upstream of the dams (Parkinson and others, 2003), as does Dworshak Dam (completed in 1972) on the North Fork Clearwater River. The Hells Canyon Complex and Dworshak Dam effectively eliminate about 81 and 25 percent, respectively, of the total drainage basin in the Snake and Clearwater Rivers from contributing sediment to Lower Granite Reservoir. The Salmon River Basin (drainage area of about 14,000 mi^2) is about 71 percent of the remaining area in the Snake River Basin downstream of Hells Canyon Dam and about 52 percent of the total combined area of the Snake and Clearwater River Basins contributing sediment to Lower Granite Reservoir. Other contributing basins are the Clearwater (about 7,200 mi^2 or 27 percent of the total area), the Grande Ronde (about 3,950 mi^2 or 15 percent), and the Snake River Basin downstream of Hells Canyon Dam to Lower Granite Reservoir, but excluding the Salmon and Grande Ronde Basins (about 1,850 mi^2 or 7 percent). The Palouse River has a drainage basin of about 3,280 mi^2, and is a tributary to the Snake River about 48 mi downstream of Lower Granite Dam.

Salmon River Basin

The Salmon River is an unregulated, free-flowing river that originates in mountain ranges in Idaho and western Montana and flows about 410 mi through central Idaho to its confluence with the Snake River in lower Hells Canyon (fig. 1). The Salmon River derives its streamflow from several tributaries including the Lemhi, Pahsimeroi, Middle Fork Salmon, South Fork Salmon, and Little Salmon Rivers. Peak flows in the Salmon River generally occur in May and June during snowmelt runoff. Between 1975 and 2010, the Salmon River, on average, discharged about 10,700 ft^3/s to the Snake River. The Salmon River contributes about 32 percent of the mean streamflow measured at the Snake River near Anatone (streamgage 13334300 [fig. 2]) and about 22 percent of the combined streamflow entering Lower Granite Reservoir from the Snake and Clearwater Rivers (U.S. Geological Survey, 2013).

About 90 percent of the Salmon River Basin is comprised of Federal lands, about 77 percent national forest managed by the U.S. Forest Service, and 13 percent managed by the Bureau of Land Management. Nearly 80 percent of the land cover is forest in the Salmon River Basin (fig. 2); agricultural and urban areas combined account for less than 3 percent (Tetra Tech, 2006). Key geologic features in the Salmon River Basin are the Idaho Batholith and Challis volcanics that tend to produce coarse, sandy soils that are highly erodible when weathered (King and others, 2004). The erodible geology, steep topography, and lack of hydrologic control structures combine to mobilize large quantities of sediment from the Salmon River Basin to downstream waters. Numerous forest fires that burned large areas of the basin between 1980 and 2010 have further worsened the susceptibility of the Salmon River Basin to erosion.

Snake River Downstream of Hells Canyon Dam

The Snake River in the Hells Canyon reach incorporates drainages upstream of Lower Granite Reservoir and downstream of Hells Canyon Dam, exclusive of the Salmon and Grande Ronde River Basins (fig. 1). The supply of sediment to this reach of the Snake River is limited by upstream trapping in the Hells Canyon Complex of dams that was completed in 1967. Although sediment delivery to the Snake River in the Hells Canyon reach is limited, streamflow is generated from the entire 93,500-mi^2 drainage basin, including the area upstream of the Hells Canyon complex. In the 110-mi reach of the Snake River between Hells Canyon Dam and Lower Granite Reservoir, the Snake River flows primarily north, forming first the border between Idaho and Oregon, and then the border between Idaho and Washington.

The primary tributary in this reach, excluding the Salmon and the Grande Ronde Rivers, is the Imnaha River which enters from the west about 55 mi downstream of Hells Canyon Dam. Between Hells Canyon Dam and the Imnaha, the Snake River runs through a deep and narrow v-shaped valley entrenched in erosion-resistant basalt and metamorphic bedrock (Parkinson and others, 2003). About 22 percent of the 1,850 mi^2 area drained by the Snake River in the Hells Canyon reach is comprised of agricultural land, primarily in the lower parts of the drainage basin downstream of the confluence with the Imnaha. Forested land makes up about 47 percent of the drainage basin (fig. 2) (Tetra Tech, 2006). Sediment source studies in the Hells Canyon reach indicate that sediments transported in the Snake River are primarily coarse grained, gravel-sized materials derived from local landslides, talus slopes, and tributary inflows (Miller and others, 2003).

Grande Ronde River Basin

The Grande Ronde River flows about 180 mi in a northeasterly direction draining streams in the Blue and Wallowa Mountains of northeastern Oregon. Most of the Grande Ronde drainage basin is in Oregon with a small part in southeastern Washington. A major tributary to the Grande Ronde River is the Wallowa River. The Grande Ronde River flows into the Snake River about 23 mi upstream of the town of Asotin, Washington, and just upstream of the sampling location on the Snake River near Anatone (fig. 2). Because most of the drainage basin in the lower Grande Ronde is at an altitude less than 3,000 ft, runoff typically starts in April and lasts until late June when snowmelt occurs in higher elevation areas of the basin.

About 70 percent of the Grande Ronde Basin is forested and about 17 percent is agricultural land (fig. 2). About 50 percent of the basin is privately owned; most of these private lands lie within stream valleys along the Grande Ronde and Wallowa Rivers. Private lands primarily are used for agriculture, grazing, and forestry (Tetra Tech, 2006). The U.S. Forest Service manages about 47 percent of the land within the Grande Ronde Basin for multiple uses including timber production, livestock grazing, and recreation. The surface geology in the Grande Ronde Basin is primarily basalt from the Columbia River Group with a highly variable soil erosional capacity. In contrast to the Salmon River Basin, where most of the sediment is derived from natural processes, land-use practices probably account for a large part, if not most of the sediment production in the Grande Ronde Basin.

Clearwater River Basin

The Clearwater River originates in the Bitterroot Mountains at the border of Idaho and Montana and flows westward to its confluence with the Snake River at Lewiston, Idaho (fig. 1). Major tributaries to the Clearwater River include the Lochsa and Selway Rivers, the North and South Forks of the Clearwater River, and the Potlatch River. Dworshak Dam effectively traps most of the sediment transported from the North Fork Clearwater River drainage basin prior to reaching the main stem Clearwater River. From 1975 through 2010, the Clearwater River contributed about 30 percent of the streamflow (as measured at Spalding) entering Lower Granite Reservoir. Similar to the Salmon River, streamflow in the Clearwater River typically peaks in May and June in response to snowmelt runoff. Highly erosive igneous rocks underlay a large part of the Clearwater River Basin (King and others, 2004). As a result, much of the basin is highly susceptible to erosion and subsequent sediment transport. Overall, cropland and pastureland make up about 18 percent of the Clearwater River Basin (Tetra Tech, 2006). The intensity of agricultural activity generally increases in a downstream or westerly direction (fig. 2).

The Lochsa and Selway Rivers combined drain about 46 percent of the Clearwater River Basin, draining areas that are essentially 100 percent forested; the U.S Forest Service manages more than 95 percent of the land (Tetra Tech, 2006). Underlain by the Idaho Batholith, the Lochsa and Selway drainage basins are characterized by rock that weathers deeply to produce coarse, sandy soils with high erosion rates if disturbed (King and others, 2004).

The Potlatch River is the largest tributary (drainage basin of about 550 mi^2) to the lower Clearwater River Basin. The Potlatch River is a tributary to the Clearwater River about 15 mi upstream of Lower Granite Reservoir (fig. 1). About 57 percent of the Potlatch River drainage basin is forested, mostly in the northern upstream areas. Primary land-uses in the forested areas include timber harvest and other forest management practices. The downstream part of the drainage basin is predominantly agricultural (about 43 percent of the total basin area), used primarily for dryland agriculture and grazing (Latah Soil and Water Conservation District, 2007). Land-use activities in the drainage basin have resulted in changes in the vegetative cover, increases in soil compaction, and channel modifications that have resulted in a flashy hydrograph and rapid streamflow runoff (Latah Soil and Water Conservation District, 2007). Instantaneous streamflows of 8,000 ft^3/s in winter and early spring are not unusual.

Palouse River Basin

The Palouse River drains about 3,300 mi^2 of southeastern Washington and parts of the northern Idaho panhandle. The headwaters of the Palouse River originate in the forested mountains of northwestern Idaho; the river flows westward through farmland to its confluence with the Snake River about 48 mi downstream of Lower Granite Dam (fig. 1). Major tributaries to the Palouse River are the South Fork Palouse River and Paradise, Rebel Flat, Rock, Union Flat, and Cow Creeks. Activities that affect water quality in the Palouse River Basin include dryland agriculture (67 percent of the drainage basin), rangeland (26 percent), timber harvest, mining, and urban development (Washington State Department of Ecology, 2006). Irrigated farmland adjacent to the Palouse River and its tributaries comprises less than 1 percent of the land use. Forested land comprises about 6 percent of the drainage basin, primarily in the upland northern and eastern parts (fig. 2). Agricultural fields throughout the basin are highly susceptible to soil erosion from November through March when high intensity rainstorms can cause intensive runoff and soil erosion. These winter storms can deliver large quantities of sediment to streams throughout the Palouse River drainage basin (Ebbert and Roe, 1998).

Methods of Data Collection and Analysis

Data were collected for this study from March 2008 through September 2011 to evaluate sediment transport in the lower Snake and Clearwater River Basins and sediment deposition in Lower Granite Reservoir. Data collection included measurement of streamflow, conventional sampling for suspended sediment and bedload at 12 sampling stations in the lower Snake and Clearwater River Basins, and acoustic surrogate sampling at two USGS streamgages: the Snake River near Anatone, Washington and the Clearwater River at Spalding, Idaho (table 1). The methods and results of depositional surveys conducted in Lower Granite Reservoir are described in Williams and others (2012) and Braun and others (2012), respectively. This section of the report documents the methods used for data collection and analysis of streamflow and suspended sediment and bedload.

Table 1. U.S. Geological Survey streamgaging stations where sediment samples were collected, lower Snake and Clearwater River Basins, Washington and Idaho, water years 2008–11.

[Locations of stations are shown in figure 2. **Effective drainage area** does not include watershed area upstream of Dworshak and Hells Canyon Dams. **Abbreviations:** USGS, U.S. Geological Survey; mi^2, square mile]

USGS gaging station No.	Gaging station name	Effective drainage area (mi^2)	Sampling period	Type of streamflow record	Number of suspended-sediment samples	Number of bedload samples
13317000	Salmon River at White Bird, Idaho	13,500	03-2009–07-2011	Continuous	42	5
13334000	Grande Ronde River at Zindel, Washington	3,940	03-2009–07-2011	Indexed	38	0
13334300	Snake River near Anatone, Washington[1]	19,700	04-2008–07-2011	Continuous	39	16
13336500	Selway River near Lowell, Idaho	1,910	03-2009–07-2011	Continuous	37	11
13337000	Lochsa River near Lowell, Idaho	1,180	03-2009–07-2011	Continuous	36	3
13337120	Middle Fork Clearwater River at Kooskia, Idaho	5,490	03-2009–07-2011	Indexed	38	11
13338100	South Fork Clearwater River near Harpster, Idaho	865	03-2009–07-2011	Indexed	39	15
13338500	South Fork Clearwater River at Stites, Idaho	1,150	03-2009–07-2011	Continuous	41	14
13340000	Clearwater River at Orofino, Idaho	5,580	03-2009–07-2011	Continuous	36	7
13341570	Potlatch River below Little Potlatch Creek near Spalding, Idaho	583	03-2009–07-2011	Continuous	37	0
13342500	Clearwater River at Spalding, Idaho[1]	7,140	03-2008–07-2011	Continuous	40	11
13351000	Palouse River at Hooper, Washington	2,500	03-2009–07-2011	Continuous	33	0

[1] Station was equipped with suspended sediment surrogate technology.

Streamflow

Streamflow measurements at all of the sediment sampling stations were obtained using standard USGS methods (Mueller and Wagner, 2009; Turnipseed and Sauer, 2010). At 9 of the 12 sampling stations, streamflow was measured at an established streamgage using a continuous record of water stage calibrated to periodic onsite measurements of streamflow. For the three sampling stations without continuous streamflow record (table 1), streamflow was measured when sediment samples were collected, and then indexed to the streamflow at a nearby streamgage (or group of streamgages) with continuous data. The correlation between streamflow at the non-continuous sampling station and the continuous streamgage(s) was used to generate a daily mean streamflow record for the non-continuous sampling station over the period of this study. For example, instantaneous streamflow measurements made during sediment sample collection at the non-continuous sampling station on the South Fork Clearwater River near Harpster, Idaho (USGS station 13338100) from March 2009 through July 31, 2011, were indexed to the streamflow recorded simultaneously at the continuous streamgage on the South Fork Clearwater River at Stites, Idaho (USGS streamgage 13338500) ($R^2 = 0.996$). The resultant relation was used to estimate the daily mean streamflow at the South Fork Clearwater River near Harpster from the daily mean streamflow at South Fork Clearwater River at Stites for the period of study. The same methodology was used for two other non-continuous sampling stations; the Grande Ronde River at Zindel, Washington (USGS station 13334000) was indexed to an upstream streamgage with continuous record on the Grande Ronde River at Troy, Oregon ($R^2 = 0.976$), and the Middle Fork Clearwater River at Kooskia, Idaho (USGS station 13337120) was indexed to the combined streamflow from the continuous streamgage on the Selway River near Lowell, Idaho (USGS streamgage 13336500) and the Lochsa River near Lowell, Idaho (USGS streamgage 13337000) ($R^2 = 0.992$).

Suspended Sediment and Bedload

Streams transport sediment by carrying the finer particles in suspension with turbulent eddies and by rolling or skipping coarser particles along the streambed. The discharge of fine-grained particles typically is controlled by the available supply of fine-grained sediment. These fine-grained sediments generally move downstream in suspension at about the same velocity as the water. The sediment that moves on or near the stream bottom by sliding, rolling, or bouncing is bedload (Edwards and Glysson, 1999), which increases exponentially with increasing streamflow. The total sediment load in a stream is defined as the sum of the suspended-sediment load plus the bedload. Because the particle size distribution of the suspended load is a function of streamflow, substantial variation typically occurs in the concentration and grain-size characteristics of sediments both spatially at different locations in a stream and temporally with changes in the magnitude of streamflow.

For this study, suspended sediment and bedload sampling started in spring 2008 at the Snake River near Anatone and the Clearwater River at Spalding streamgages. In spring 2009, sample collection extended to 10 additional stations for suspended sediment and 7 stations for bedload to help quantify the sediment contributions from discrete subbasins in the lower Snake and Clearwater River Basins (table 1). Sample collection at all the stations continued through July 2011, following the end of the spring runoff. All sediment samples were collected using standard USGS methods, procedures, and equipment as documented by Edwards and Glysson (1999).

Suspended Sediment

Suspended-sediment samples were collected 9–15 times per year at each of the study stations to encompass a range of streamflow conditions (table 2). As illustrated in figure 3 at the Snake River near Anatone and the Clearwater River at Spalding, targeted suspended-sediment samples were collected during the ascending limb, the peak, and the descending limb of the snowmelt runoff hydrograph at each station to determine the effects of hysteresis on SSC. Significant hydrologic events such as a rain-on-snow event that occurred during January 2011 were targeted for additional sampling.

Suspended-sediment samples were collected using equal-width and depth-integrating techniques. Isokinetic D-96 collapsible-bag samplers were suspended from either a bridge or cableway to obtain the samples. Use of the D-96 allows for isokinetic sampling over a wide range of stream depths and velocities (Davis, 2005). For collection of suspended-sediment samples, the total stream width at each station was divided into 10 equal-width increments, and individual depth-integrated samples were collected at the centroid of each increment.

Table 2. Ranges of streamflow sampled and suspended-sediment concentrations in samples collected from the lower Snake and Clearwater River Basins, Washington and Idaho, water years 2008–11.

[Locations of stations are shown in figure 2. **Abbreviations:** USGS, U.S. Geological Survey; ft³/s, cubic foot per second; mg/L, milligram per liter; μm, micrometer]

USGS gaging station No.	Gaging station name	Streamflow range when samples collected (ft³/s)	Median suspended-sediment concentration (mg/L)	Range of suspended-sediment concentration (mg/L)	Median suspended-sediment fraction less than 62.5 μm (percent)	Range suspended-sediment concentration less than 62.5 μm (percent)
13317000	Salmon River at White Bird, Idaho	3,970–83,900	94	3.0–600	51	14–97
13334000	Grande Ronde River at Zindel, Washington	766–25,000	30	3.0–570	84	65–97
13334300	Snake River near Anatone, Washington	14,900–155,000	55	5.0–410	62	32–94
13336500	Selway River near Lowell, Idaho	651–32,500	11	1.0–180	42	15–89
13337000	Lochsa River near Lowell, Idaho	434–21,200	11	1.0–87	51	23–84
13337120	Middle Fork Clearwater River at Kooskia, Idaho	1,140–28,000	15	1.0–99	49	23–94
13338100	South Fork Clearwater River near Harpster, Idaho	194–11,000	26	1.0–200	61	28–90
13338500	South Fork Clearwater River at Stites, Idaho	213–12,400	24	2.0–530	62	39–94
13340000	Clearwater River at Orofino, Idaho	1,580–69,500	13	2.0–240	66	29–91
13341570	Potlatch River below Little Potlatch Creek near Spalding, Idaho	15.0–17,000	42	3.0–3,300	92	68–99
13342500	Clearwater River at Spalding, Idaho	3,390–78,900	21	3.0–210	70	30–96
13351000	Palouse River at Hooper, Washington	48.0–7,150	100	12–1,400	95	48–99

Individual samples from each centroid were composited in 3-L bottles until the entire cross section of the stream was sampled. Typically, two–three 3-L bottles were required for each suspended-sediment composite. Samples were sent to the USGS Cascade Volcano Observatory (CVO) sediment laboratory in Vancouver, Washington after collection where they were analyzed for total SSC and particle-size fraction less than 62.5 μm (Guy, 1969). In addition to the composited samples, selected suspended-sediment samples were collected at the Snake River near Anatone and Clearwater River at Spalding streamgages to document the river cross-sectional variability. For these samples, individual bottles were used to collect samples from each centroid in the cross section. The samples were then analyzed separately for sediment concentration and full-grain size analysis on the sediment size greater than a diameter of 62.5 μm.

The SSC in a stream generally varies in relation to streamflow. Because of this variability, summary statistics to characterize a stream, such as a mean concentration, may include bias resulting from variation in the sampling frequency and the timing of sampling over the stream hydrograph or a storm event. For this study, mean flow-weighted concentrations and suspended-sediment loads were simulated using LOADEST, a FORTRAN program for estimating constituent loads in streams and rivers (Runkel and others, 2004). LOADEST is based on a rating-curve method (Cohn and others, 1989, 1992; Crawford, 1991) that uses regression to estimate constituent load in relation to several predictor variables related to streamflow and time. This type of model has been used to estimate constituent concentrations for periods when sample data were not available (Gilroy and others, 1990), to estimate a basin flux of water-quality constituents (Goolsby and others, 1999), and to evaluate long-term trends in water-quality data (Smith and others, 1987).

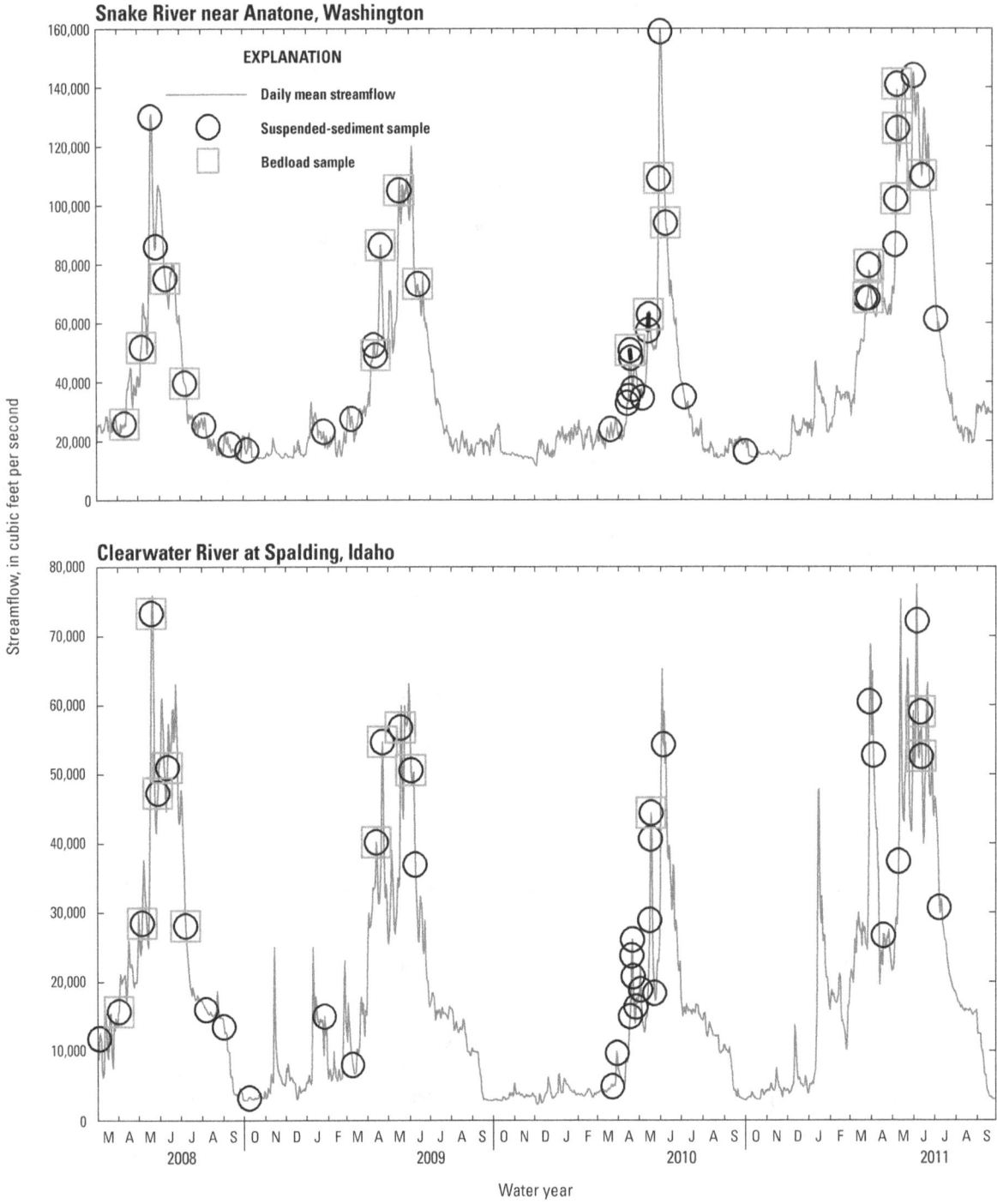

Figure 3. Streamflows at which suspended-sediment and bedload samples were collected from the Snake River near Anatone, Washington, and Clearwater River at Spalding, Idaho, water years 2008–11.

For this study, LOADEST was used to develop regression models for estimating the loads and flow-weighted concentrations of total suspended sediment (TSS) and the sand and fine-grained fractions of suspended sediment for each station for water years 2009 through 2011. For the streamgages on the Snake River near Anatone and the Clearwater River at Spalding, loads and concentrations were estimated for water years 2008 through 2011. The regression model used for the TSS and suspended-sediment size fractions was:

$$\ln L = I + a(\ln Q) + b(\ln Q^2) + c[\sin(2\pi T)] \quad (1)$$
$$+ d[\cos(2\pi T)] + \varepsilon$$

where

L is the suspended-sediment load, in pounds per day;

I is the regression intercept;

Q is the centered streamflow, in cubic feet per second;

T is the centered decimal time in years from the beginning of the calibration period;

$a, b, c,$ and d are regression coefficients that remain constant over time; and

ε is unaccounted error associated with the regression model.

For each model, the predictor variables in the regression equation were selected on the basis of Aikaike Information Criteria (Aikaike, 1981; Judge and others, 1985). The criteria are designed to achieve a good compromise between using as many predictor variables as possible to explain the variance in load while minimizing the standard error of the resulting estimates. Estimates of the daily constituent load for each station were computed using the selected model (table 3) and daily mean streamflow. Bias introduced by conversion of the logarithm of load into estimates of actual load were corrected using the Bradu-Mundlak method (Bradu and Mundlak, 1970; Cohn and others, 1989; Crawford, 1991). A mean flow-weighted concentration for suspended sediment at each station was estimated as the TSS load over a given time period divided by the total streamflow at the station during the same period.

Suspended-Sediment Surrogates

The need to measure fluvial sediment has led to the development of sediment surrogate technologies, particularly in locations where streamflow alone may not be a good estimator of SSC because of regulated flow, hysteresis effect on sediment concentration, episodic sediment sources, and non-equilibrium sediment transport. An effective surrogate technology is low-maintenance and robust over a range of hydrologic conditions, and measured variables can be modeled to SSC, sediment load, and duration of elevated levels on a real-time and continuous basis. Although numerous surrogate technologies were tested during this study (including turbidity and laser diffraction), only acoustic backscatter measurements using ADVMs is described in this report. Wood and Teasdale (2013) present a detailed discussion for each of the surrogate technologies tested during this study, and compare and contrast the results.

Acoustic backscatter has been used successfully as a surrogate for SSC in the San Francisco Bay, California (Gartner, 2004), estuaries in Florida (Patino and Byrne, 2004), Colorado River (Topping and others, 2004), Hudson River, New York (Wall and others, 2006), the Aegean Region, Turkey (Elci and others, 2009), and subtropical estuaries in Australia (Chanson and others, 2008). For the acoustic backscatter method, a device is used that emits an acoustic pulse into the water. Theoretically, the strength of the reflected pulse, called backscatter, increases when more particles, in this case assumed to be sediment, are present in the water. Acoustic instruments have shown great promise in sediment surrogate studies because they are robust to biological fouling and measure profiles of backscatter across a sampling volume rather than at a single point in the stream (Gartner and Gray, 2005). Acoustic frequencies were selected for this study to maximize sensitivity of backscatter to dominant grain size (lower frequency for the sand fraction and higher frequency for the fines fraction) and to minimize errors due to changing grain-size distribution, as described in Gartner (2004) and Topping and others (2004). To estimate the concentrations of total suspended sediment, sands, and fines, acoustic backscatter data were corrected for (1) beam spreading, (2) transmission losses due to absorption by water, and (3) absorption or attenuation by sediment. In this study, the Clearwater River at Spalding and the Snake River near Anatone streamgages (table 1) were instrumented with surrogate technologies.

The equal-width-increment suspended-sediment samples collected at the lower Snake and Clearwater River gaging stations were used to develop the relation between SSC and ADVM backscatter. The sampling strategy targeted the ascending limb, the peak, and the descending limb of the snowmelt runoff hydrograph at each station. Forty suspended-sediment samples from the Clearwater River and 39 samples from the lower Snake River were collected during the study period March 2008 to July 2011. Samples submitted for analysis were a composite representative of the entire cross section. As mentioned previously, cross-sectional variability in SSC in the Clearwater and Snake Rivers was evaluated by individually analyzing 10 separate vertically integrated samples from the width of the river channel. Cross-section variability was analyzed for four sample sets collected from the Clearwater River and five sample sets collected from the Snake River. All suspended-sediment samples collected from the Snake River near Anatone and Clearwater River at Spalding streamgages were analyzed for organic content using a loss-on ignition analysis (Schumacher, 2002) to evaluate the effect of organic matter on errors in the surrogate analysis.

Table 3. Regression coefficients and coefficients of determination for models used to estimate concentrations and loads of total suspended sediment and suspended-sediment size fractions at stations in the lower Snake and Clearwater River Basins, Washington and Idaho.

[Locations of stations are shown in figure 2. The regression equation is $\ln L = I + a(\ln Q) + b(\ln Q^2) + c[\sin(2\pi T)] + d[\cos(2\pi T)] + \varepsilon$, where L is the sediment load in kilograms per day; I is the regression intercept; Q is the centered stream discharge in cubic feet per second; T is the centered decimal time in years from the beginning of the calibration period; a, b, c, and d, are regression coefficients that remain constant over time; ε is unaccounted error associated with the regression model; and R^2 (coefficient of determination) represents the amount of variance explained by the model. USGS, U.S. Geological Survey]

USGS gaging station No.	Gaging station name	I	Regression coefficient				R^2 (percent)
			a	b	c	d	
Total suspended sediment							
13317000	Salmon River at White Bird, Idaho	15.47	2.518	-0.079	-0.684	-0.790	89
13334000	Grande Ronde River at Zindel, Washington	11.84	2.236	0.575	-0.713	0.506	91
13334300	Snake River near Anatone, Washington	15.69	3.093	0.084	-0.540	0.160	95
13336500	Selway River near Lowell, Idaho	11.02	2.609	0.512	-0.923	-0.437	95
13337000	Lochsa River near Lowell, Idaho	10.38	2.193	0.402	-0.948	0.024	96
13337120	Middle Fork Clearwater River at Kooskia, Idaho	11.86	2.250	0.402	-0.947	-0.308	95
13338100	South Fork Clearwater River near Harpster, Idaho	12.04	2.714	0.030	-0.796	-1.610	96
13338500	South Fork Clearwater River at Stites, Idaho	11.04	2.463	0.328	-0.730	-0.442	95
13340000	Clearwater River at Orofino, Idaho	12.24	2.229	0.630	-0.831	-0.311	93
13341570	Potlatch River below Little Potlatch Creek near Spalding, Idaho	11.08	2.251	0.126	-0.143	-0.697	93
13342500	Clearwater River at Spalding, Idaho	12.57	2.047	0.590	0.461	-0.477	90
13351000	Palouse River at Hooper, Washington	12.64	1.655	0.021	-0.930	-0.210	87
Total suspended sands							
13317000	Salmon River at White Bird, Idaho	14.51	3.212	-0.197	-1.098	-0.816	95
13334000	Grande Ronde River at Zindel, Washington	9.97	2.797	0.847	-0.541	-0.134	92
13334300	Snake River near Anatone, Washington	14.70	3.993	-0.465	-0.490	-0.056	97
13336500	Selway River near Lowell, Idaho	9.32	2.798	0.659	-0.755	0.394	95
13337000	Lochsa River near Lowell, Idaho	9.84	2.588	0.414	-0.798	-0.582	96
13337120	Middle Fork Clearwater River at Kooskia, Idaho	10.95	2.696	0.490	-0.558	-0.670	96
13338100	South Fork Clearwater River near Harpster, Idaho	11.30	3.190	-0.114	-0.459	-1.867	96
13338500	South Fork Clearwater River at Stites, Idaho	9.95	2.889	0.396	-0.301	-0.709	97
13340000	Clearwater River at Orofino, Idaho	10.65	2.587	0.751	-0.607	-0.003	95
13341570	Potlatch River below Little Potlatch, near Spalding, Idaho	8.49	2.156	0.191	-0.327	-0.982	97
13342500	Clearwater River at Spalding, Idaho	10.83	2.525	0.694	0.242	-0.854	92
13351000	Palouse River at Hooper, Washington	9.54	1.376	0.068	-0.727	0.148	69
Total suspended fines							
13317000	Salmon River at White Bird, Idaho	15.20	2.349	-0.076	-0.831	-1.281	82
13334000	Grande Ronde River at Zindel, Washington	11.65	2.150	0.507	-0.709	0.607	90
13334300	Snake River near Anatone, Washington	15.12	2.717	0.225	-0.563	0.342	92
13336500	Selway River near Lowell, Idaho	10.89	2.377	0.321	-1.033	-0.933	92
13337000	Lochsa River near Lowell, Idaho	9.76	1.978	0.350	-1.099	0.148	93
13337120	Middle Fork Clearwater River at Kooskia, Idaho	11.28	1.981	0.353	-1.171	-0.236	91
13338100	South Fork Clearwater River near Harpster, Idaho	11.41	2.469	0.063	-0.977	-1.538	94
13338500	South Fork Clearwater River at Stites, Idaho	10.53	2.250	0.292	-0.857	-0.281	93
13340000	Clearwater River at Orofino, Idaho	11.92	1.991	0.545	-0.907	-0.376	90
13341570	Potlatch River below Little Potlatch, near Spalding, Idaho	10.95	2.270	0.120	-0.131	-0.658	92
13342500	Clearwater River at Spalding, Idaho	12.30	1.845	0.518	0.651	-0.358	86
13351000	Palouse River at Hooper, Washington	12.56	1.694	0.021	-0.928	-0.265	88

The ADVMs at the Clearwater River at Spalding were co-located with the streamgage and installed about 16 ft upstream of the streamgage. In May 2008, the station was equipped with a 0.5 megahertz (MHz) Sontek™ Argonaut-SL ADVM and 3 MHz Sontek™ Argonaut-SL ADVM with datalogger and satellite telemetry. The ADVMs were mounted on an aluminum slide track that could be raised and lowered as needed to service equipment. The 0.5 and 3 MHz ADVMs measured backscatter in five discrete, equally sized cells in a horizontal sampling volume, 5.0–100 ft and 3.3–12 ft from the instrument, respectively. The sampling volume for each ADVM was selected based on meter frequency, availability of suspended material to reflect the acoustic pulse, and any obstructions in the beam path. The ADVMs collected backscatter data 2 minutes of every 15 minutes.

In April 2009, the Snake River near Anatone streamgage was equipped with a 0.5 MHz Sontek Argonaut-SL ADVM and 1.5 MHz Argonaut-SL ADVM with datalogger and satellite telemetry. The station was located on the left streambank about 1,000 ft upstream of the streamgage and the sediment sampling location. The 0.5 and 1.5 MHz ADVMs were mounted on an aluminum slide track and were configured to measure backscatter in five discrete, equally sized cells in a horizontal sampling volume, 6.6–203 and 6.6–59 ft from the instrument, respectively. The ADVMs collected backscatter data 2 minutes of every 15 minutes.

Acoustic backscatter data must be corrected prior to relating the data to the mass concentration of suspended sediment in the water column. Factors affecting acoustic backscatter readings include transmission losses due to absorption by water, absorption or attenuation by sediment, and beam spreading. However, the methods for correcting backscatter data differ and can substantially alter estimates of SSC. Thus, selection of an appropriate method is important in the analysis of acoustic backscatter data. For this study, candidate methods for correcting the backscatter data were reviewed; methods selected for use with data collected at the Snake River and Clearwater River stations are documented in Wood and Teasdale (2013).

Following correction of the backscatter data, measurements were averaged over a 1-hour period bracketing the time that sediment samples were collected. This allowed concurrent measurements to relate backscatter data to the measured SSC at each station. Relations between backscatter and SSC were evaluated using ordinary least squares regression. All of the SSC data were log transformed prior to regression. Acoustic backscatter data are reported in a log-based scale and do not require a transformation. Regression models were selected based on statistical significance of explanatory variables (p-values) and various regression statistics such as coefficient of determination (R^2), standard error, Mallow's Cp (Ott and Longnecker, 2001), and prediction error sum of squares (PRESS) (Helsel and Hirsch, 1992) statistics. A nonparametric bias correction factor described in Duan (1983) was applied to each best-fit regression model to correct for low bias induced by log transformation and subsequent re-transformation of the dependent variable. The factor was used to correct each value of SSC as well as upper and lower 95 percent confidence intervals estimated by a regression model.

Bedload

Unlike suspended-sediment transport, bedload often is not detectable. When bedload transport does occur, it is often extremely variable both spatially within the stream channel and temporally during steady streamflow conditions (Hubbell, 1964). During this study, bedload was collected at most of the stations when streamflow was sufficiently high to initiate bedload transport. However, at some stations, the stream cross-section was not conducive for collecting bedload because of poor channel geometry, backwater conditions, extreme turbulence, or a combination of these factors. Bedload samples were collected at only two of the three stations in the Snake River drainage basin upstream of lower Granite Reservoir. The channel bottom at the Grande Ronde River at Zindel, Washington, sampling station is composed of uneven bedrock, making the site nearly impossible to sample with the Helley-Smith bedload sampler. In the Clearwater River Basin, 7 of the 8 stations were sampled at least 3 times for bedload, with the Lochsa River at Lowell limited to 3 samples because of backwater conditions from the Selway River. The Potlatch River was not sampled because of backwater conditions from the Clearwater River. Although bedload sampling was attempted on numerous occasions at the Palouse River station, bedload was never retained in the sampler. It is difficult to determine whether the lack of bedload in the Palouse River at Hooper was real or an artifact of a poor sampling location. The number of bedload samples collected at each station is listed in table 1. The timing of bedload sampling for the Snake River near Anatone and the Clearwater River at Spalding streamgages from March 2008 through July 2011 is shown in figure 3.

Most bedload samples were collected using a Helley-Smith type sampler with a 6- by 6-in. nozzle (Helley and Smith, 1971); a few samples were collected in 2011 using an Elwha style sampler with a 4- by 8-in. nozzle. An equal-width-increment sampling method as described by Edwards and Glysson (1999) was used to collect bedload. This method involves dividing and collecting samples in the stream channel at evenly spaced sections to represent bedload transport accurately across the entire channel. When possible, samples were collected at 20 equally spaced intervals with duration of 60 seconds on the stream bottom per interval. At some stations and at some streamflows, fewer than 20 sections were sampled because of extremely fast and turbulent stream conditions.

For analysis, all sectional samples were composited into one sample representing the entire cross section. Duplicates were collected during each sampling event and analyzed separately at the USGS CVO sediment laboratory. For analysis, the results for the duplicate samples are reported as an average. The bedload samples were analyzed for total mass and full phi (ϕ) particle size increments from $\phi = 4$ (62.5 µm) to $\phi = -7$ (128 mm).

Because of the large variability in the spatial and temporal pattern of bedload transport, curves relating bedload to streamflow generally have a larger degree of uncertainty than do curves for suspended-sediment transport. To estimate bedload for use in transport curves, the following equation (Edwards and Glysson, 1999), was used:

$$Qb = (k \times W \times M) / T \qquad (2)$$

where

Qb is bedload discharge, in tons per day;

k is a conversion factor specific to the sampler orifice (0.381 for the Helley Smith and 0.141 for the Elwha);

W is the total stream width, in feet;

M is the total mass of bedload sample, in grams and

T is the total time the sampler was on the bed, in seconds.

The best-fit equation between streamflow and bedload discharge was used to estimate the total bedload discharge for the period of interest using either the continuous or indexed daily streamflow record. Because of the paucity of samples at some stations and the variability at most of the stations, the bedload discharge estimates obtained during this study have a wide range of associated error and should not be viewed as precise estimates.

The grain size and statistics software GRADISTAT (Blott and Pye, 2001) was used to classify the particle-size distribution data. Emmett (1976) described the particle size distribution as bimodal at the Snake River near Anatone and the Clearwater River at Spalding streamgages. Standard statistical measures such as the arithmetic (normal distribution) or geometric (log-normal) method of moments are not suitable for identifying the bimodal particle size distribution because of non-normality. Folk and Ward (1957) described a graphical geometric (log-normal) method to determine grain-size analysis statistics. Blott and Pye (2001) determined this method to be the most appropriate for parameter-based estimation of bimodal bedload transport. Parameters used to describe the particle-size distribution in this study included sample distribution (unimodal, bimodal), cumulative exceedence values (D_{10}, D_{50}, D_{90}), and cumulative percentage finer by weight. Standard deviation

(sorting), skewness (positive or negative preference), and kurtosis (peakedness around the mean) were not used in this analysis as they were determined to be unreliable for bimodal bedload distributions.

Sediment Transport in the Lower Snake and Clearwater River Basins

Because sediment transport in streams varies in response to hydrologic conditions, information on streamflow in a historical context is critical for assessing sediment transport and deposition in the lower Snake and Clearwater River Basins. The magnitude and timing of streamflow in the lower Snake and Clearwater River Basins generally are determined by the amount of water derived from the winter snowpack and reservoir operations at Hells Canyon and Dworshak Reservoirs. As such, rivers and streams in the lower Snake and Clearwater River Basins typically reach peak flows in April, May, or June in association with spring snowmelt runoff. High streamflows occasionally occur in lower elevation basins in the autumn and winter in association with rain-on-snow events that can cause flooding and substantial transport of sediment. Generally, however, when precipitation and snowpack are below average, streamflow from runoff is lower than normal, and transport of sediment in streams is below average. In contrast, when precipitation, snowpack, and runoff are above average, sediment transport may be larger than normal. Generally, streamflows drop rapidly over the summer following the loss of the snowpack, and low flow typically is in September or October.

From March 2008 through September 2011, streamflow conditions in the lower Snake and Clearwater River Basins as represented by the streamgages on the Snake River near Anatone and the Clearwater River at Spalding were quite variable in relation to the most recent 30-year average (fig. 4). Streamflow in the Snake and Clearwater Rivers during water years 2008, 2009, and, in particular, 2011 generally exceeded the 30-year average during most of the year. During 2010, streamflow in the Snake and Clearwater Rivers was less than the 30-year average for most of the year, exceeding the average streamflow only during the brief period of snowmelt runoff from late May through early June. A pronounced rain-on-snow event occurred during January 2011 affecting streamflows and sediment transport in the Potlatch River below Little Potlatch near Spalding (USGS streamgage 13341570) and the Palouse River at Hooper (USGS streamgage 13351000). This event occurred across many of the lower-elevation drainage basins in the lower Snake and Clearwater River Basins and is evident in the hydrograph spike in late January 2011 for the Snake River near Anatone and in particular for the Clearwater River at Spalding (fig. 4).

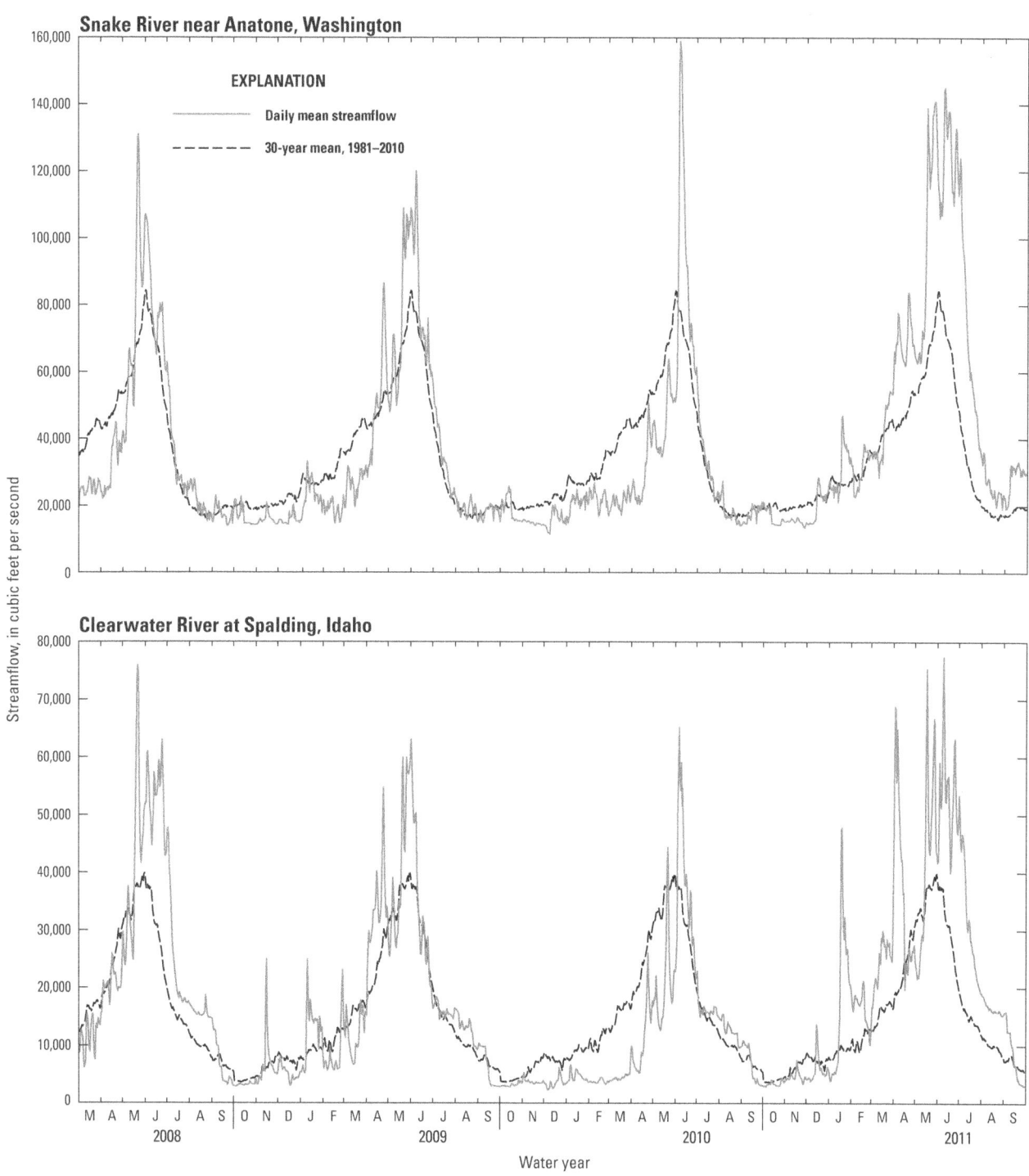

Figure 4. Daily mean streamflow in the Snake River near Anatone, Washington and the Clearwater River at Spalding, Idaho, 2008–11, compared with the 30-year mean for water years 1981–2010.

Suspended-Sediment Transport

The number of suspended-sediment samples collected varied from 33 to 42 per station during the sampling period (table 1). Fitted regression lines for a log-log power fit with associated R^2 values of TSS and streamflow for selected stations in the lower Snake and Clearwater River Basins are shown in figure 5. Because there is considerable scatter around the regression lines and the R^2 values at some stations were relatively low, a degree of uncertainty exists in the TSS load calculations based on streamflow. However, assuming that the streamflows in individual subbasins during 2009–11 are representative of normal conditions, their relative contributions to the overall sediment load delivered to Lower Granite Reservoir probably are accurate. Based on the hydrographs for 2009–11 compared to the 30-year mean for the Snake River near Anatone and the Clearwater River at Spalding (fig. 4), this assumption probably is valid.

The concentrations and load of suspended sediment transported in streams were variable, spanning 3–5 orders of magnitude during water years 2008–11 (fig. 6). The largest measured concentrations of suspended sediment were in samples collected from the Palouse and Potlatch Rivers during the rain-on-snow event in January 2011. Concentrations of TSS, which includes both the sand and fine-grained fractions, in the Potlatch River ranged more than 4 orders of magnitude during the study from a low of 3 mg/L during baseflow conditions to more than 3,000 mg/L in January 2011 during the rain-on-snow event (table 2). Samples collected from the Palouse River at Hooper had the largest median concentration of TSS (100 mg/L) and the largest suspended sediment fraction as fine-grained silt and clay less than 62.5 μm in diameter (median 95 percent). Other rivers with a large percentage of fine-grained suspended sediment were the Potlatch River (median 92 percent) and the Grand Ronde River (median 84 percent) (table 2). In contrast to the other stations sampled during this study, the Palouse and Potlatch Rivers and, to a lesser extent, the Grande Ronde River showed an increase in the fraction of fine-grained sediment with increasing streamflow. The Palouse, Potlatch, and Grande Ronde Rivers all drain basins with relatively large proportions of agricultural activity, which probably accounts for the relatively large percentage of fine-grained sediment.

The second largest median TSS concentration (94 mg/L) was measured in samples collected at Salmon River at White Bird, Idaho (USGS streamgage 13317000). However, as a percentage of the TSS, the fine-grained fraction (median of 51 percent) was smaller in samples collected in the Salmon River as compared with the Palouse, Potlatch, and Grande Ronde Rivers. The smallest TSS with median concentrations of 11, 11, 15, and 13 mg/L were measured in samples collected from the Selway and Lochsa Rivers, the Middle Fork Clearwater River at Kooskia, and the Clearwater River at Orofino (USGS streamgage 13340000), respectively. The TSS concentration in samples from the Lochsa River and the Middle Fork Clearwater River at Kooskia did not exceed 100 mg/L in any of the samples collected during this study (table 2).

Samples collected from the Selway River, Lochsa River, and the Middle Fork Clearwater River at Kooskia, and samples from the Salmon River at White Bird, had relatively small percentages of fine-grained sediment. The median silt and clay fractions at each of these stations were about 50 percent or less. All stations in the network, except the agriculturally affected Palouse, Potlatch, and Grande Ronde Rivers, exhibited an increase in the sand-size fraction of suspended sediment with an increase in streamflow. As a result, during the peak of spring snowmelt runoff, most of the suspended sediment transported in the lower Snake and Clearwater River Basins was sand-sized material larger than 62.5 μm.

Based on the transport curves (fig. 6B), the load of TSS in the Palouse River ranged from about 1 to more than 10,000 tons per day and in the Potlatch River ranged from less than 0.1 to more 40,000 tons per day. However, because the transport curves in figure 6B represent a best-fit line through the data for each station, during peak runoff the actual concentration and load of TSS is often much larger than the load estimated using the sediment-transport curves. This is evident in the stations shown in figure 5 where the best-fit regression line for each station underestimates the measured TSS concentration at high streamflow. For instance, multiple samples collected from the Palouse and Potlatch Rivers during the January 2011 rain-on-snow event indicate that the suspended-sediment transport far exceeded the estimates based on the transport curves shown in figure 6. For the Palouse River, four TSS samples collected January 17–18 at streamflows ranging from 5,670 to 7,150 ft^3/s had concentrations between 930 and 1,400 mg/L, about twice the amount estimated using the transport curve values of 490–600 mg/L, respectively. For the Potlatch River, five samples collected January 16–17 at streamflows ranging from 8,760 to 17,000 ft^3/s had TSS concentrations of between 1,400 and 3,300 mg/L, more than three times the estimated concentrations of 420 and 940 mg/L, respectively.

Because the LOADEST program uses multiple variables to predict SSC and load, simulation results from the model generally provide better estimates than those based on transport curves as shown in figure 6. LOADEST was therefore used to develop regression models (table 3) for each station to estimate the load of TSS, total suspended sand, and total suspended fines for water years 2009–11(table 4). To determine the annual load, the daily load for each constituent was estimated using the daily mean streamflow and summed to determine the annual load. Regression coefficients and coefficients of determination (R^2) for the LOADEST models at each station are given in table 3.

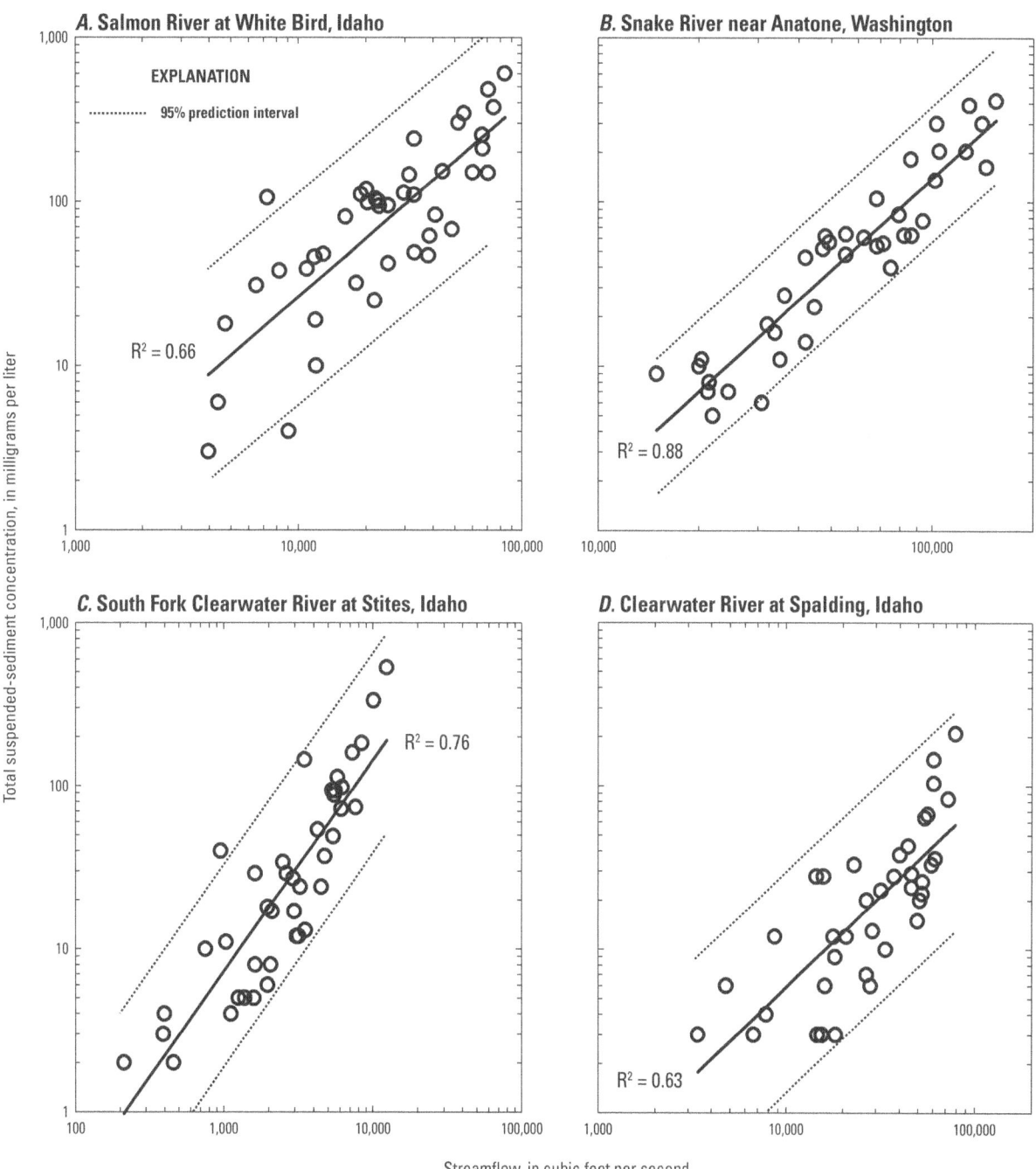

Figure 5. Suspended-sediment transport curves and 95-percent prediction intervals representing best-fit regression equations for a log-log power fit of total suspended sediment with streamflow in the (*A*) Salmon River at Whitebird, Idaho, (*B*) Snake River near Anatone, Washington, (*C*) South Fork Clearwater River at Stites, Idaho, and (*D*) Clearwater River at Spalding, Idaho.

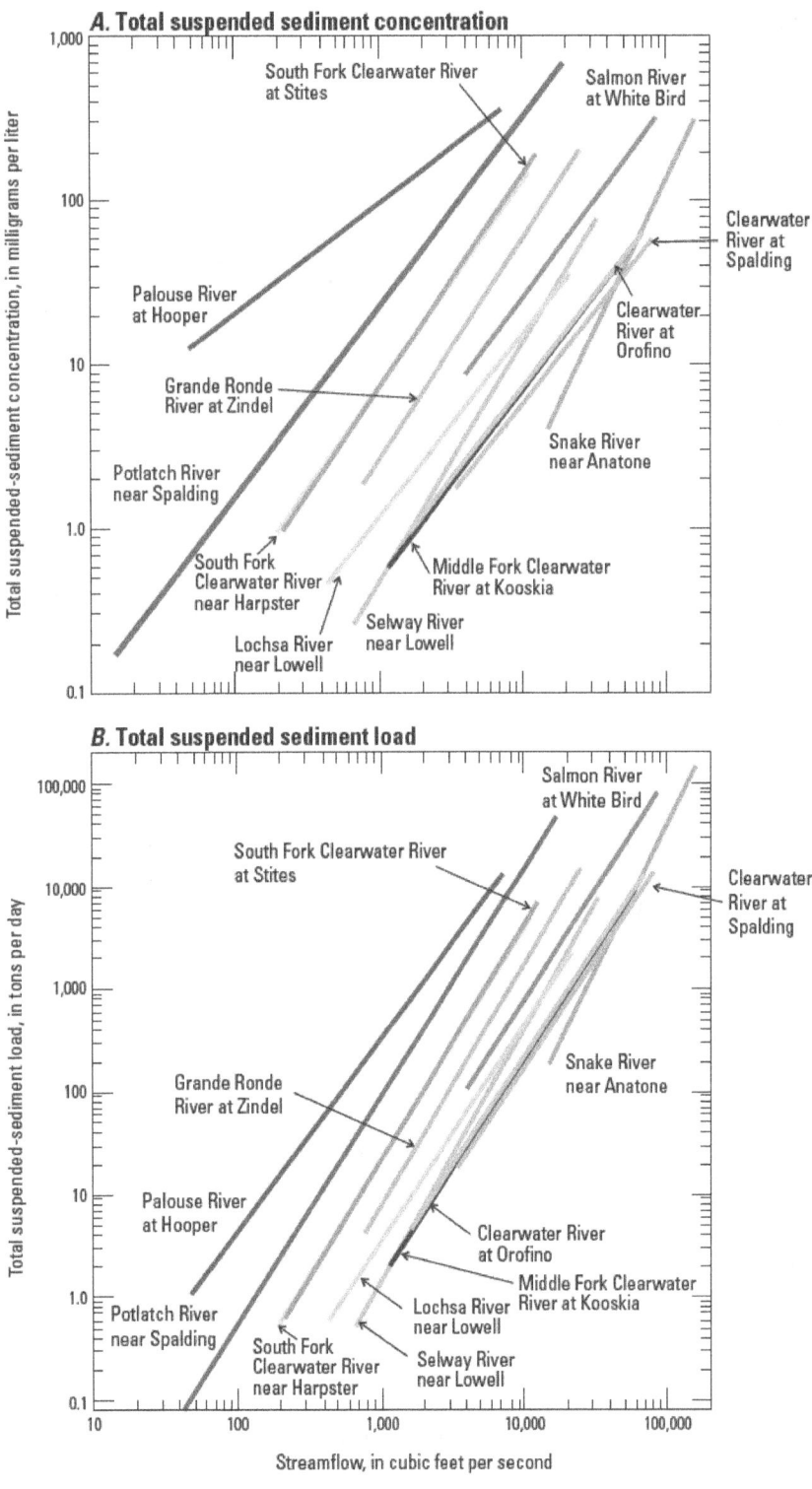

Figure 6.   Suspended-sediment transport curves representing best-fit regression equations for (A) total suspended-sediment concentrations and (B) total suspended-sediment loads at sampling stations in the lower Snake and Clearwater River Basins, Washington and Idaho.

High R^2 values indicate that the LOADEST models did an excellent job of accounting for the variability in the suspended- sediment loads: 30 of the 36 models accounted for 90 percent or more of observed variability. The models account for less than 80 percent of the variability in the observed sediment load only for the Palouse River and only for the sand-size sediment fraction. Generally, the model simulation results showed the best fit (highest R^2) for the sand size fraction and the poorest fit for the fine-grained fraction. A notable exception was for the Palouse River, where the best-fit model occurred for total suspended fines ($R^2 = 0.88$) and the poorest fit occurred for the total suspended sands ($R^2 = 0.69$).

For each of the stations in the sampling network, the estimated loads of total suspended sediment (fig. 7), suspended sands, and suspended fines were largest during water year 2011 and smallest during water year 2010 (table 4). The difference in the TSS load for the Palouse and Potlatch Rivers during water years 2010 and 2011 was particularly notable. The TSS load in the Palouse River was about 8 times larger in 2011 than in 2010 and in the Potlatch River was more than 50 times larger in 2011 than in 2010. The large difference in the sediment load between 2011 and 2010 in the Palouse and Potlatch Rivers is attributable to the rain-on-snow event that occurred in these drainage basins during January 2011. In the Palouse River at Hooper, the TSS load during January 17–19, 2011, was about 64.600 tons, roughly 19 percent of the 2011 water year total of 335,000 tons and about 9 percent of the TSS load transported in the Palouse River during water years 2009–11 (table 4). During January 16–17, 2011, the estimated TSS load in the Potlatch River exceeded 110,000 tons, more than one-half of the total estimated load of 212,000 tons for the entire 2011 water year and about 40 percent of the total TSS load transported in the Potlatch River during water years 2009–11 (table 4).

Table 4.   Estimated annual, total, and mean loads of suspended sediment, suspended sand, and suspended fines for stations in the lower Snake and Clearwater River Basins, Washington and Idaho, water years 2009–11.

[Locations of stations are shown in figure 2. **Abbreviations:** C.I , confidence interval; USGS, U.S. Geological Survey]

USGS gaging station No.	Gaging station name	Total suspended sediment (tons)			Total suspended sands (tons)			Total suspended fines (tons)		
		Estimated load	Lower 95 percent C.I.	Upper 95 percent C.I.	Estimated load	Lower 95 percent C.I.	Upper 95 percent C.I.	Estimated load	Lower 95 percent C.I.	Upper 95 percent C.I.
Water year 2009										
13317000	Salmon River at White Bird, Idaho	1,660,000	962,000	2,720,000	952,000	590,000	1,450,000	744,000	387,000	1,330,000
13334000	Grande Ronde River at Zindel, Washington	278,000	174,000	425,000	52,200	29,500	86,900	224,000	138,000	347,000
13334300	Snake River near Anatone, Washington	1,820,000	1,340,000	2,420,000	848,000	631,000	1,110,000	998,000	687,000	1,410,000
13336500	Selway River near Lowell, Idaho	153,000	87,700	250,000	116,000	59,600	203,000	46,600	25,400	81,400
13337000	Lochsa River near Lowell, Idaho	42,300	29,000	60,700	24,500	15,300	38,200	18,700	12,000	28,500
13337120	Middle Fork Clearwater River at Kooskia, Idaho	194,000	127,000	289,000	115,000	72,500	175,000	86,000	50,300	141,000
13338100	South Fork Clearwater River near Harpster, Idaho	44,600	31,200	63,200	22,100	14,500	32,900	22,700	15,000	34,100
13338500	South Fork Clearwater River at Stites, Idaho	48,800	33,400	69,400	18,700	13,200	26,100	28,900	18,600	43,100
13340000	Clearwater River at Orofino, Idaho	300,000	182,000	475,000	154,000	89,200	249,000	150,000	85,500	252,000
13341570	Potlatch River below Little Potlatch Creek, near Spalding, Idaho	57,800	20,600	135,000	2,840	1,460	5,120	55,500	18,600	134,000
13342500	Clearwater River at Spalding, Idaho	335,000	215,000	501,000	137,000	81,600	215,000	199,000	122,000	310,000
13351000	Palouse River at Hooper, Washington	318,000	130,000	672,000	19,600	4,950	55,500	290,000	123,000	596,000

Table 4.   Estimated annual, total, and mean loads of suspended sediment, suspended sand, and suspended fines for stations in the lower Snake and Clearwater River Basins, Washington and Idaho, water years 2009–11.—Continued

[Locations of stations are shown in figure 2. **Abbreviations:** C.I., confidence interval; USGS, U.S. Geological Survey]

USGS gaging station No.	Gaging station name	Total suspended sediment (tons)			Total suspended sands (tons)			Total suspended fines (tons)		
		Estimated load	Lower 95 percent C.I.	Upper 95 percent C.I.	Estimated load	Lower 95 percent C.I.	Upper 95 percent C.I.	Estimated load	Lower 95 percent C.I.	Upper 95 percent C.I.
					Water Year 2010					
13317000	Salmon River at White Bird, Idaho	1,030,000	590,000	1,700,000	524,000	321,000	811,000	476,000	242,000	876,000
13334000	Grande Ronde River at Zindel, Washington	144,000	73,600	258,000	46,700	18,300	99,600	109,000	55,300	195,000
13334300	Snake River near Anatone, Washington	1,650,000	1,130,000	2,320,000	795,000	552,000	1,110,000	855,000	543,000	1,290,000
13336500	Selway River near Lowell, Idaho	60,700	32,500	104,000	44,800	20,400	86,000	19,400	10,300	33,600
13337000	Lochsa River near Lowell, Idaho	18,300	12,700	25,800	9,720	6,120	14,800	8,810	5,660	13,200
13337120	Middle Fork Clearwater River at Kooskia, Idaho	81,600	53,000	121,000	44,800	27,900	68,600	38,800	22,400	63,400
13338100	South Fork Clearwater River near Harpster, Idaho	14,200	9,860	19,900	7,440	4,800	11,000	7,260	4,730	10,800
13338500	South Fork Clearwater River at Stites, Idaho	16,200	10,700	23,500	6,360	4,340	8,990	9,850	6,120	15,100
13340000	Clearwater River at Orofino, Idaho	143,000	82,900	232,000	71,400	38,900	121,000	75,100	40,800	130,000
13341570	Potlatch River below Little Potlatch Creek near Spalding, Idaho	4,090	982	11,700	229	93	482	3,810	844	11,300
13342500	Clearwater River at Spalding, Idaho	152,000	92,200	237,000	58,900	33,500	96,700	91,200	52,200	150,000
13351000	Palouse River at Hooper, Washington	39,800	12,700	96,700	2,960	572	9,410	35,000	11,800	82,500

Table 4. Estimated annual, total, and mean loads of suspended sediment, suspended sand, and suspended fines for stations in the lower Snake and Clearwater River Basins, Washington and Idaho, water years 2009–11.—Continued

[Locations of stations are shown in figure 2. **Abbreviations:** C.I , confidence interval; USGS, U.S. Geological Survey]

USGS gaging station No.	Gaging station name	Total suspended sediment (tons)			Total suspended sands (tons)			Total suspended fines (tons)		
		Estimated load	Lower 95 percent C.I.	Upper 95 percent C.I.	Estimated load	Lower 95 percent C.I.	Upper 95 percent C.I.	Estimated load	Lower 95 percent C.I.	Upper 95 percent C.I.
				Water year 2011						
13317000	Salmon River at White Bird, Idaho	2,400,000	1,410,000	3,870,000	1,300,000	832,000	1,950,000	1,060,000	560,000	1,870,000
13334000	Grande Ronde River at Zindel, Washington	288,000	164,000	489,000	60,500	28,600	122,000	228,000	129,000	389,000
13334300	Snake River near Anatone, Washington	5,450,000	3,860,000	7,500,000	2,800,000	2,000,000	3,800,000	2,700,000	1,770,000	3,970,000
13336500	Selway River near Lowell, Idaho	154,000	91,500	247,000	109,000	58,500	186,000	50,500	28,200	86,300
13337000	Lochsa River near Lowell, Idaho	65,200	44,500	93,600	40,400	25,400	62,100	27,800	17,500	42,900
13337120	Middle Fork Clearwater River at Kooskia, Idaho	239,000	159,000	351,000	147,000	95,700	217,000	104,000	60,900	170,000
13338100	South Fork Clearwater River near Harpster, Idaho	87,100	53,300	142,000	46,500	26,500	79,300	41,700	23,800	72,800
13338500	South Fork Clearwater River at Stites, Idaho	101,000	60,600	159,000	51,300	31,000	80,900	52,900	30,200	87,200
13340000	Clearwater River at Orofino, Idaho	517,000	301,000	845,000	285,000	159,000	479,000	243,000	132,000	423,000
13341570	Potlatch River below Little Potlatch Creek near Spalding, Idaho	212,000	50,900	607,000	12,900	5,090	27,400	199,000	44,700	598,000
13342500	Clearwater River at Spalding, Idaho	662,000	394,000	1,050,000	297,000	164,000	498,000	372,000	210,000	618,000
13351000	Palouse River at Hooper, Washington	335,000	139,000	691,000	20,900	5,210	58,600	304,000	132,000	608,000

Table 4.   Estimated annual, total, and mean loads of suspended sediment, suspended sand, and suspended fines for stations in the lower Snake and Clearwater River Basins, Washington and Idaho, water years 2009–11.—Continued

[Locations of stations are shown in figure 2. **Abbreviations:** C.I., confidence interval; USGS, U.S. Geological Survey]

USGS gaging station No.	Gaging station name	Total suspended sediment (tons)			Total suspended sands (tons)			Total suspended fines (tons)		
		Estimated load	Lower 95 percent C.I.	Upper 95 percent C.I.	Estimated load	Lower 95 percent C.I.	Upper 95 percent C.I.	Estimated load	Lower 95 percent C.I.	Upper 95 percent C.I.
					Water years 2009–11					
13317000	Salmon River at White Bird, Idaho	5,090,000	2,960,000	8,290,000	2,780,000	1,740,000	4,210,000	2,280,000	1,190,000	4,080,000
13334000	Grande Ronde River at Zindel, Washington	710,000	412,000	1,170,000	159,400	76,400	308,000	561,000	322,000	931,000
13334300	Snake River near Anatone, Washington	8,920,000	6,330,000	12,200,000	4,440,000	3,180,000	6,020,000	4,550,000	3,000,000	6,670,000
13336500	Selway River near Lowell, Idaho	368,000	212,000	601,000	270,000	138,000	475,000	116,000	63,900	201,000
13337000	Lochsa River near Lowell, Idaho	126,000	86,200	180,000	74,600	46,800	115,000	55,300	35,200	84,600
13337120	Middle Fork Clearwater River at Kooskia, Idaho	515,000	339,000	761,000	307,000	196,000	461,000	229,000	134,000	374,000
13338100	South Fork Clearwater River near Harpster, Idaho	146,000	94,400	225,000	76,000	45,800	123,000	71,700	43,500	118,000
13338500	South Fork Clearwater River at Stites, Idaho	166,000	105,000	252,000	76,400	48,500	116,000	91,600	54,900	145,000
13340000	Clearwater River at Orofino, Idaho	960,000	566,000	1,550,000	510,000	287,000	849,000	468,000	258,000	805,000
13341570	Potlatch River below Little Potlatch Creek near Spalding, Idaho	274,000	72,500	754,000	16,000	6,640	33,000	258,000	64,100	743,000
13342500	Clearwater River at Spalding, Idaho	1,150,000	701,000	1,790,000	493,000	279,000	810,000	662,000	384,000	1,080,000
13351000	Palouse River at Hooper, Washington	693,000	282,000	1,460,000	43,500	10,700	124,000	629,000	267,000	1,290,000

Table 4.   Estimated annual, total, and mean loads of suspended sediment, suspended sand, and suspended fines for stations in the lower Snake and Clearwater River Basins, Washington and Idaho, water years 2009–11.—Continued

[Locations of stations are shown in figure 2. **Abbreviations:** C.I , confidence interval; USGS, U.S. Geological Survey]

USGS gaging station No.	Gaging station name	Mean suspended sediment (tons)			Mean suspended sands (tons)			Mean total suspended fines (tons)		
		Estimated load	Lower 95 percent C.I.	Upper 95 percent C.I.	Estimated load	Lower 95 percent C.I.	Upper 95 percent C.I.	Estimated load	Lower 95 percent C.I.	Upper 95 percent C.I.
	Water years 2009–11									
13317000	Salmon River at White Bird, Idaho	1,696,667	986,667	2,763,333	926,667	580,000	1,403,333	760,000	396,667	1,360,000
13334000	Grande Ronde River at Zindel, Washington	236,667	137,333	390,000	53,133	25,467	102,667	187,000	107,333	310,333
13334300	Snake River near Anatone, Washington	2,973,333	2,110,000	4,066,667	1,480,000	1,060,000	2,006,667	1,516,667	1,000,000	2,223,333
13336500	Selway River near Lowell, Idaho	122,667	70,667	200,333	90,000	46,000	158,333	38,667	21,300	67,000
13337000	Lochsa River near Lowell, Idaho	42,000	28,733	60,000	24,867	15,600	38,333	18,433	11,733	28,200
13337120	Middle Fork Clearwater River at Kooskia, Idaho	171,667	113,000	253,667	102,333	65,333	153,667	76,333	44,667	124,667
13338100	South Fork Clearwater River near Harpster, Idaho	48,667	31,467	75,000	25,333	15,267	41,000	23,900	14,500	39,333
13338500	South Fork Clearwater River at Stites, Idaho	55,333	35,000	84,000	25,467	16,167	38,667	30,533	18,300	48,333
13340000	Clearwater River at Orofino, Idaho	320,000	188,667	516,667	170,000	95,667	283,000	156,000	86,000	268,333
13341570	Potlatch River below Little Potlatch Creek near Spalding, Idaho	91,333	24,167	251,333	5,333	2,213	11,000	86,000	21,367	247,667
13342500	Clearwater River at Spalding, Idaho	383,333	233,667	596,667	164,333	93,000	270,000	220,667	128,000	360,000
13351000	Palouse River at Hooper, Washington	231,000	94,000	486,667	14,500	3,567	41,333	209,667	89,000	430,000

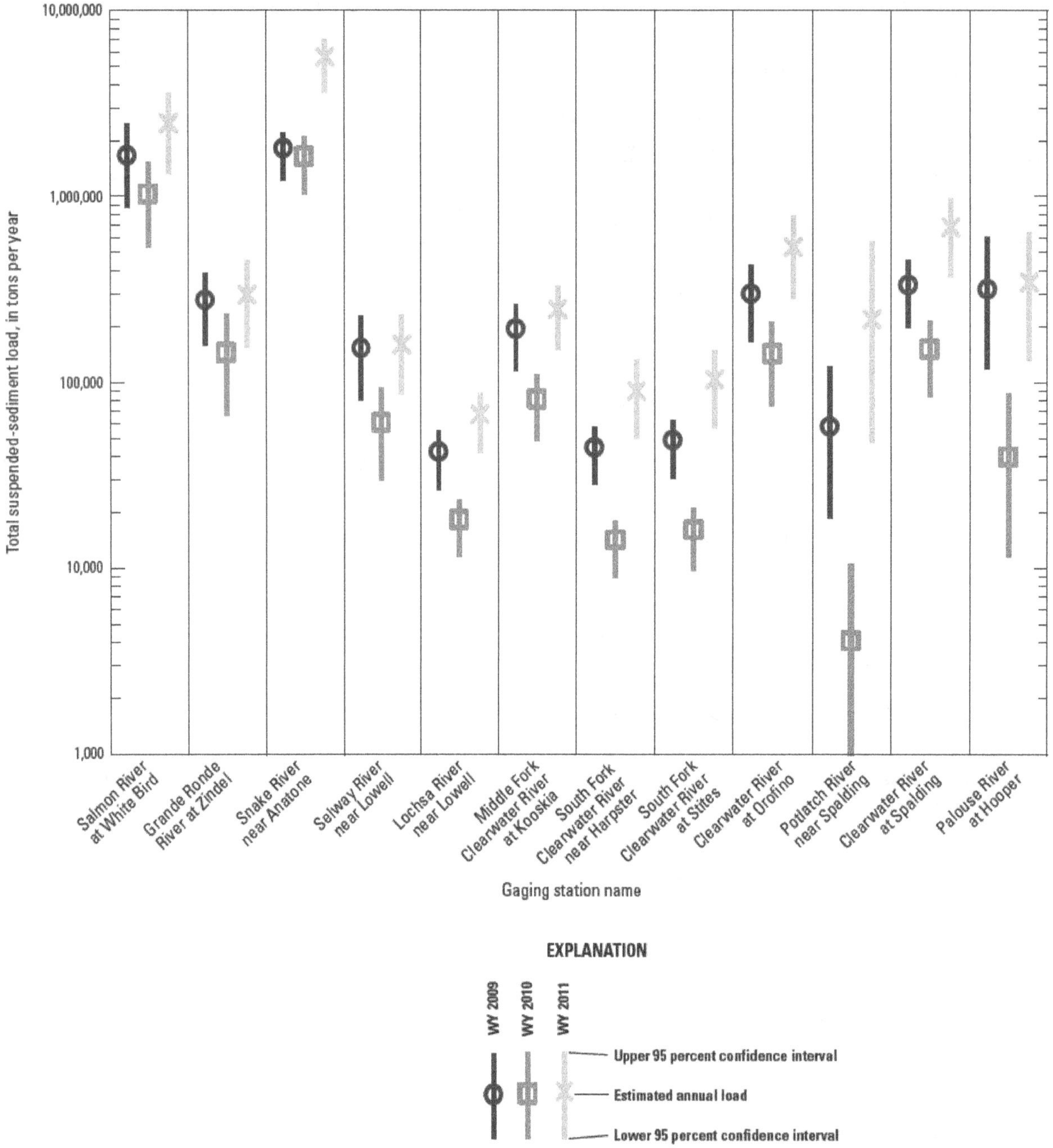

Figure 7.   Estimated total suspended-sediment loads and 95-percent confidence intervals for stations in the lower Snake and Clearwater River Basins, Idaho and Washington, water years (WY) 2009–11.

The Salmon River as measured at White Bird contributed most of the TSS load in the Snake River as measured at the Anatone station and a large part of the TSS load entering Lower Granite Reservoir. However, the relative contribution of the Salmon River to the TSS load entering the reservoir varied annually during water years 2009, 2010, and 2011. During water year 2009, the Salmon River contributed an estimated 1.66 million tons of suspended sediment to the Snake River and accounted for more than 90 percent of the 1.82 million tons of suspended sediment entering Lower Granite Reservoir from the Snake River. During water years 2010 and 2011, the Salmon River contributed 1.03 million and 2.4 million tons, or about 62 and 44 percent, respectively, of the suspended sediment entering the reservoir from the

Snake River drainage basin. The suspended-sediment yield from the Salmon River Basin, particularly the sand-sized fraction, was one of the largest of the basins evaluated (fig. 8). During the 3 years of sample collection, the mean annual yield of suspended sediment from the Salmon River Basin was about 125 tons per square mile per year [(tons/mi^2)/yr] of which about 55 percent was sand. Overall, during water years 2009–11, the Salmon River transported about 5.1 million tons of suspended sediment to the Snake River, equivalent to about 51 percent of the TSS, about 56 percent of the suspended sand, and about 44 percent of the suspended fine-grained load entering Lower Granite Reservoir from the combined Snake and Clearwater Rivers (fig. 9A–C).

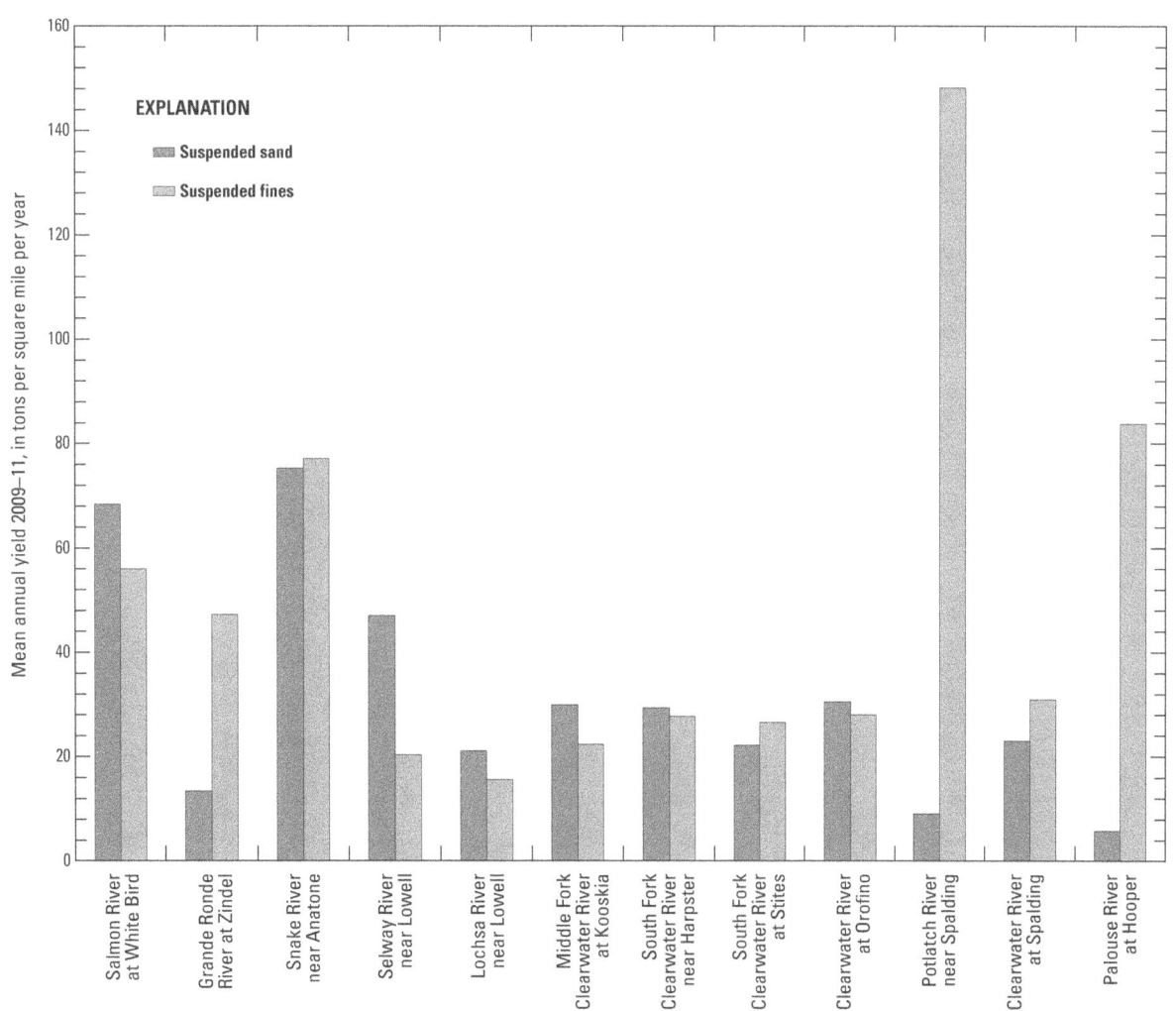

Figure 8.   Mean annual yield of suspended sand and fines in the lower Snake and Clearwater River Basins, Washington and Idaho, water years 2009–11.

A. Total suspended sediment

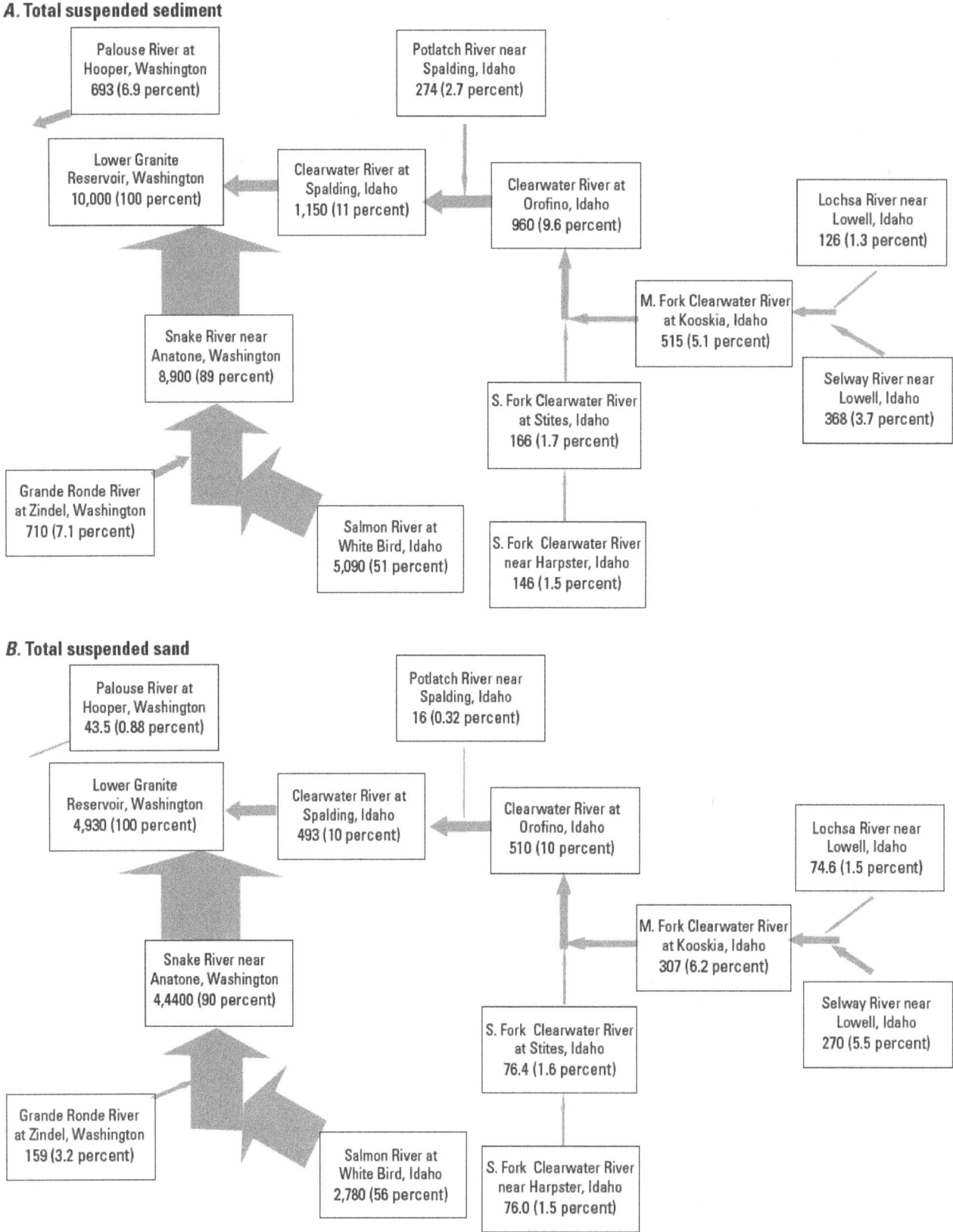

B. Total suspended sand

Figure 9.   Estimated loads of (*A*) total suspended sediment, (*B*) total suspended sand, and (*C*) total suspended fines transported in the lower Snake and Clearwater River Basins, water years 2009–11. Values are in thousands of tons and percentage of total load entering Lower Granite Reservoir during water years 2009–11. Width of each arrow is proportional to the estimated suspended sediment load.

C. Total suspended fines

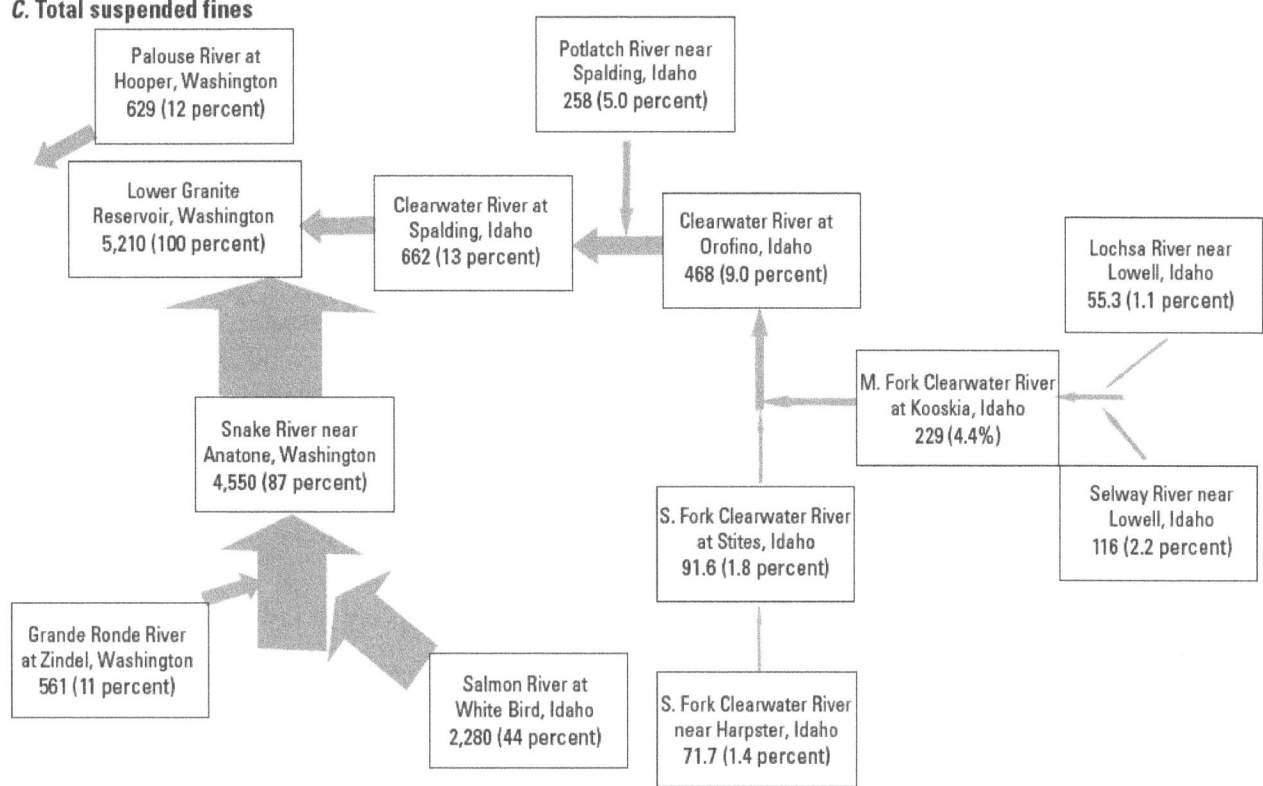

Figure 9.—Continued

The Grande Ronde River enters the Snake River just upstream of the Snake River near Anatone station (fig. 2). During water years 2009–11, the Grande Ronde River as measured at the Zindel station, contributed about 710,000 tons of suspended sediment to the Snake River (fig. 9A), about 79 percent of which was fine-grained sediment (table 4). The mean annual yield of suspended sediment from the Grande Ronde River Basin was about 60 (tons/mi^2)/yr, of which about 47 (tons/mi^2)/yr was fine-grained sediment (fig. 8). Overall, the TSS load in the Grande Ronde River was equivalent to about 7.1 percent of the suspended sediment, about 3.2 percent of the suspended sand, and about 11 percent of the suspended fines that entered Lower Granite Reservoir during water years 2009–11 (fig. 9A–C).

A primary contributor of TSS in the Clearwater drainage basin, particularly the sand-size fraction, is the Selway River, which discharged about 368,000 tons of TSS during water years 2009–11. The Selway River accounted for about 71 percent of the TSS and about 88 percent of the suspended sand as measured downstream in the Middle Fork Clearwater River at Kooskia. The Lochsa River contributed about 126,000 tons, or about 24 percent of the TSS in the Middle Fork Clearwater River at Kooskia. Of the TSS discharged from the Selway and Lochsa Rivers, about 73 and 59 percent, respectively, was sand-sized. The mean annual basin yields of suspended sediment from the Selway and Lochsa Rivers were 64 and 36 (tons/mi^2)/yr, respectively during 2009–11.

Overall, the sediment load delivered from the Selway River drainage basin was equivalent to about 32 percent of the TSS and about 55 percent of the suspended sand discharged to Lower Granite Reservoir from the Clearwater River during water years 2009–11. The TSS load from the Lochsa River drainage basin was equivalent to about 11 percent of the TSS and about 15 percent of the suspended sand discharged from the Clearwater River Basin during water years 2009–11. As a percentage of the TSS load entering Lower Granite Reservoir from both the Snake and Clearwater Rivers, the Selway River accounted for only about 3.7 and 5.5 percent, and the Lochsa River only about 1.3 and 1.5 percent of the TSS and total suspended sand, respectively (fig. 9A and 9B).

Combined, the Middle Fork Clearwater River (as measured at Kooskia, Idaho) and the South Fork Clearwater River (as measured at Stites, Idaho) discharged about 681,000 tons of suspended sediment during water years 2009–11 (fig. 9A). Of this total, about 76 percent, or 515,000 tons was from the Middle Fork Clearwater River, equivalent to about 45 percent of the TSS and 62 percent of the suspended sand entering Lower Granite Reservoir from the Clearwater River. TSS discharged from the South Fork Clearwater River was equivalent to about 14 percent of the TSS and 16 percent of the suspended sand entering Lower Granite Reservoir from the Clearwater River during water years 2009–11.

From the confluence of the South Fork Clearwater River at Stites and the Middle Fork Clearwater River at Kooskia downstream to the station at Orofino, the Clearwater River accrued about 279,000 tons of suspended sediment during water years 2009–11 (fig. 9A). Of this accrual, about 54 percent was fine-grained sediment. The Potlatch River contributed an additional 274,000 tons of suspended sediment to the Clearwater River. About 94 percent of the suspended sediment discharged from the Potlatch River was fine-grained sediment. The mean annual yield of TSS (156 [tons/mi^2]/yr) and of fine-grained suspended sediment (148 [tons/mi^2]/yr) from the Potlatch River Basin were the largest of all the stations monitored during water years 2009–11 (fig. 8). During water year 2011, the yield of TSS and fine-grained sediment from the Potlatch River Basin were about 362 and 343 (tons/mi^2)/yr, respectively. Although the TSS load from the Potlatch River was equivalent to about 24 percent of the TSS load entering Lower Granite Reservoir from the Clearwater River during water years 2009–11, the fine-grained sediment discharged from the Potlatch River was equivalent to about 39 percent of the fine-grained load entering the reservoir from the Clearwater River. During water year 2011, the load from the Potlatch River was equivalent to about 32 percent of the TSS and 53 percent of the fine-grained suspended sediment as measured in the Clearwater River at Spalding. During water year 2011, streamflow in the Potlatch River accounted for less than 1 percent of the total combined streamflow entering Lower Granite Reservoir from the Snake and Clearwater Basins. However, the TSS load and the fine-grained suspended-sediment load from the Potlatch River were equivalent to about 3.5 and 6.5 percent, respectively of the total transported to Lower Granite Reservoir, most of which was generated during the rain-on-snow event in mid-January.

The mean annual yield of TSS in the Palouse River as measured at Hooper, Washington, was about 92 (tons/mi^2)/yr during water years 2009–11. More than 90 percent of the sediment transported in the Palouse River was fine-grained sediment. Although the Palouse River discharges to the Snake River downstream of Lower Granite Reservoir, the load of suspended sediment transported in the Palouse River during water years 2009–11 was about 693,000 tons, equivalent to about 6.9 percent of the total discharged to Lower Granite Reservoir during water years 2009–11 (fig. 9A). The load of fine-grained sediment transported in the Palouse River during the same period was about 629,000 tons, equivalent to about 95 percent of the 662,000 tons of fine-grained sediment transported from the Clearwater River to Lower Granite Reservoir (fig. 9C).

Suspended Sediment Delivery to Lower Granite Reservoir

Combined, the Snake and Clearwater Rivers discharged about 10 million tons of suspended sediment to Lower Granite Reservoir during water years 2009–11 (table 4; fig. 9A). However, the delivery of sediment to Lower Granite Reservoir varied annually. About 60 percent of the 3-year total was discharged to the reservoir during water year 2011, which was characterized by a large winter snowpack, a sustained spring runoff, and high TSS loads in the Snake and Clearwater Rivers through most of the summer (fig. 10). During water years 2009–11, the Snake River accounted for about 89 percent of the TSS, about 90 percent of the suspended sand, and about 87 percent of the suspended fines entering Lower Granite Reservoir (figs. 9A–C). Only during water year 2010 did the Clearwater River account for more than 15 percent of the TSS load entering Lower Granite Reservoir.

Comparison with 1972–79 Study

Comparing data from this study with data collected during water years 1972–79 as described in Jones and Seitz (1980) indicate that, on average, the Clearwater River contributed a larger percentage of the TSS load entering Lower Granite Reservoir during 1972–79 than during 2008–11. Jones and Seitz (1980) used sediment transport curves to estimate loads, and they reported that, on an average annual basis, the Snake River accounted for about 80 percent and the Clearwater River about 20 percent of the TSS load entering Lower Granite Reservoir during 1972–79. To evaluate differences in the sediment transport characteristics between the data collected by Jones and Seitz and the data collected during this study, sediment transport curves for both study periods were compared (fig. 11). For the Snake River near Anatone (fig. 11A), the transport curves for TSS show larger concentrations and loads over the entire range of streamflow during 2008–11 as compared to 1972–79. An Analysis of Covariance (ANCOVA) (Helsel and Hirsch, 1992) indicates that the difference in concentrations of TSS (p=0.006, fig. 11A) and suspended sands (p<0.001, fig. 12B) were significantly larger (α=0.05) during 2008–11 as compared to 1972–79. However, the concentrations of suspended fines (p=0.193) were not significantly different (fig. 12A). The best-fit lines on figure 11B for the Clearwater River at Spalding indicate that the concentrations and load of suspended sediment in relation to streamflow was similar during 1972–79 as compared to 2008–11. An ANCOVA

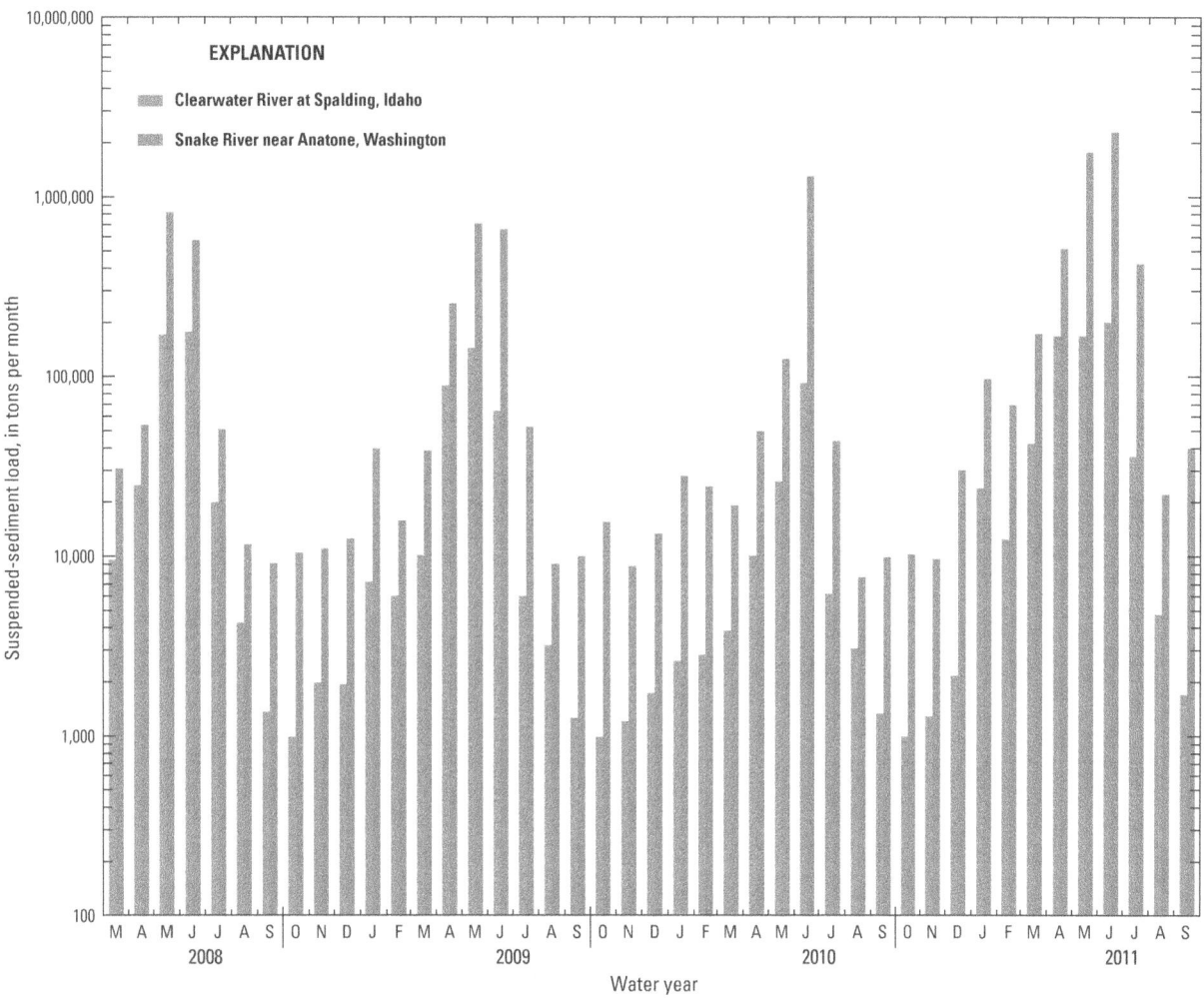

Figure 10. Suspended-sediment loads delivered monthly to Lower Granite Reservoir from the Snake and Clearwater Rivers, Washington and Idaho, March 2008–September 2011.

indicates that the difference in concentrations of TSS (p=0.853) and suspended fines (p=0.666) in the Clearwater River were not statistically significant between the two periods. However, the suspended-sand concentration in the Clearwater River at Spalding was significantly larger (p=0.011) during 2008–11 as compared to 1972–79.

Using continuous streamflow records and suspended-sediment data collected during 1972–79, loads for suspended sand and suspended fines were estimated for the Snake River near Anatone and the Clearwater River at Spalding using the LOADEST model. The suspended sand and suspended fines were analyzed separately for the 1970s data to estimate the fractional loads during each year for 1972–79 and 2009–11 (fig. 13). The results indicate that the TSS load entering Lower

Granite Reservoir from the Snake River increased from an annual average of about 71 percent of the total in water years 1972–79 to 89 percent in water years 2009–11. Conversely, the load from the Clearwater River decreased from 29 percent during 1972–79 to 11 percent during 2009–11.

As a proportion of the TSS load entering Lower Granite Reservoir from the combined Snake and Clearwater Rivers, the sand fraction increased from an annual average of about 30 percent during 1972–79 to 48 percent during 2009–11. Most of the increase in the sand load was attributable to the Snake River. In the Snake River near Anatone, the sand fraction increased from an average of 28 percent of the TSS load during 1972–79 to an average of 48 percent during 2009–11 (fig. 13). Data for the Salmon River are not sufficient to estimate suspended-sediment loads for 1972–79.

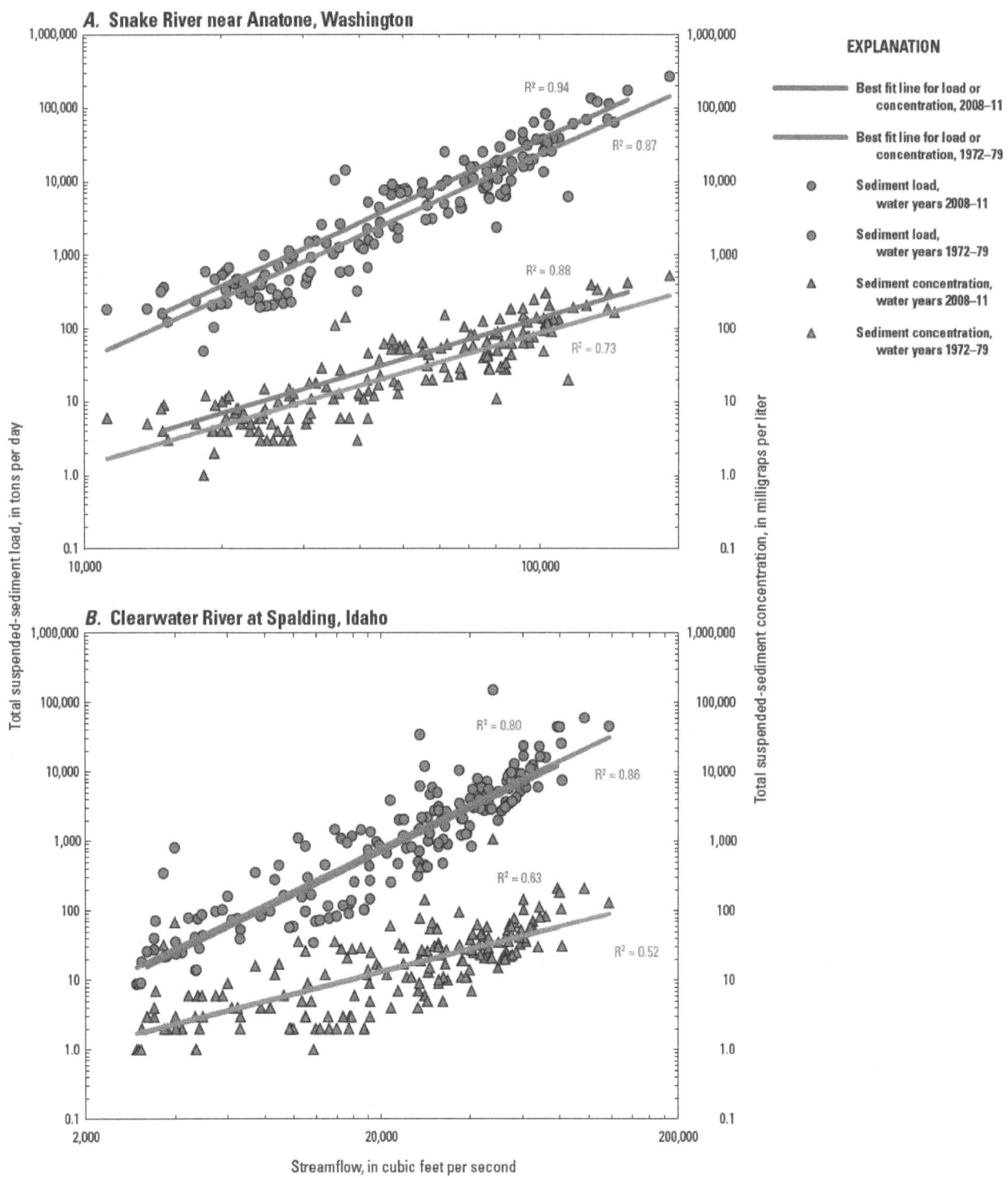

Figure 11. Suspended-sediment transport curves for concentrations and loads in the (*A*) Snake River near Anatone, Washington, and (*B*) Clearwater River at Spalding, Idaho, for data collected during water years 1972–79 and 2008–11.

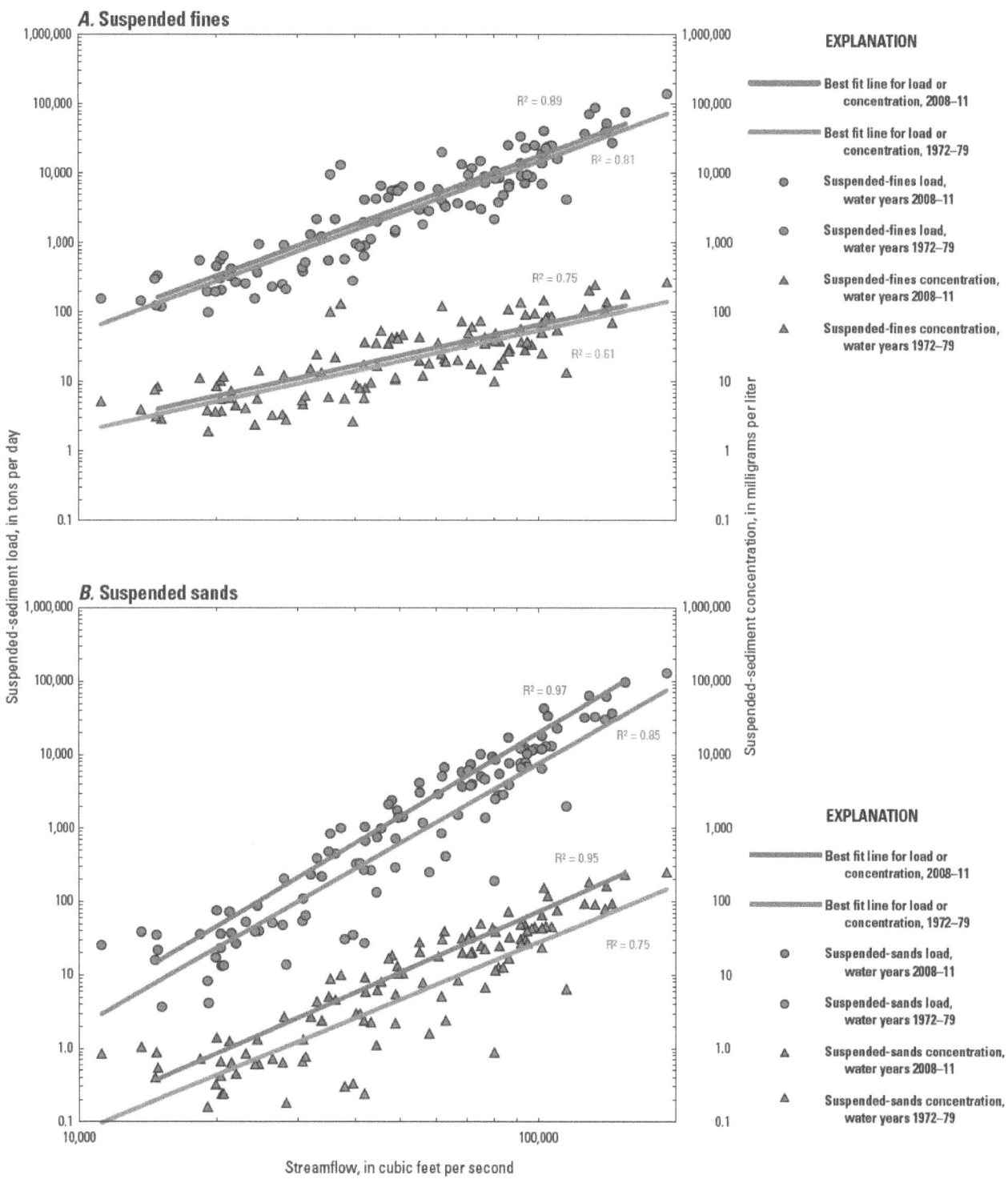

Figure 12. Suspended-sediment transport curves for concentrations and loads for (*A*) suspended fines and (*B*) suspended sands, Snake River near Anatone, Washington, water years 1972–79 and 2008–11.

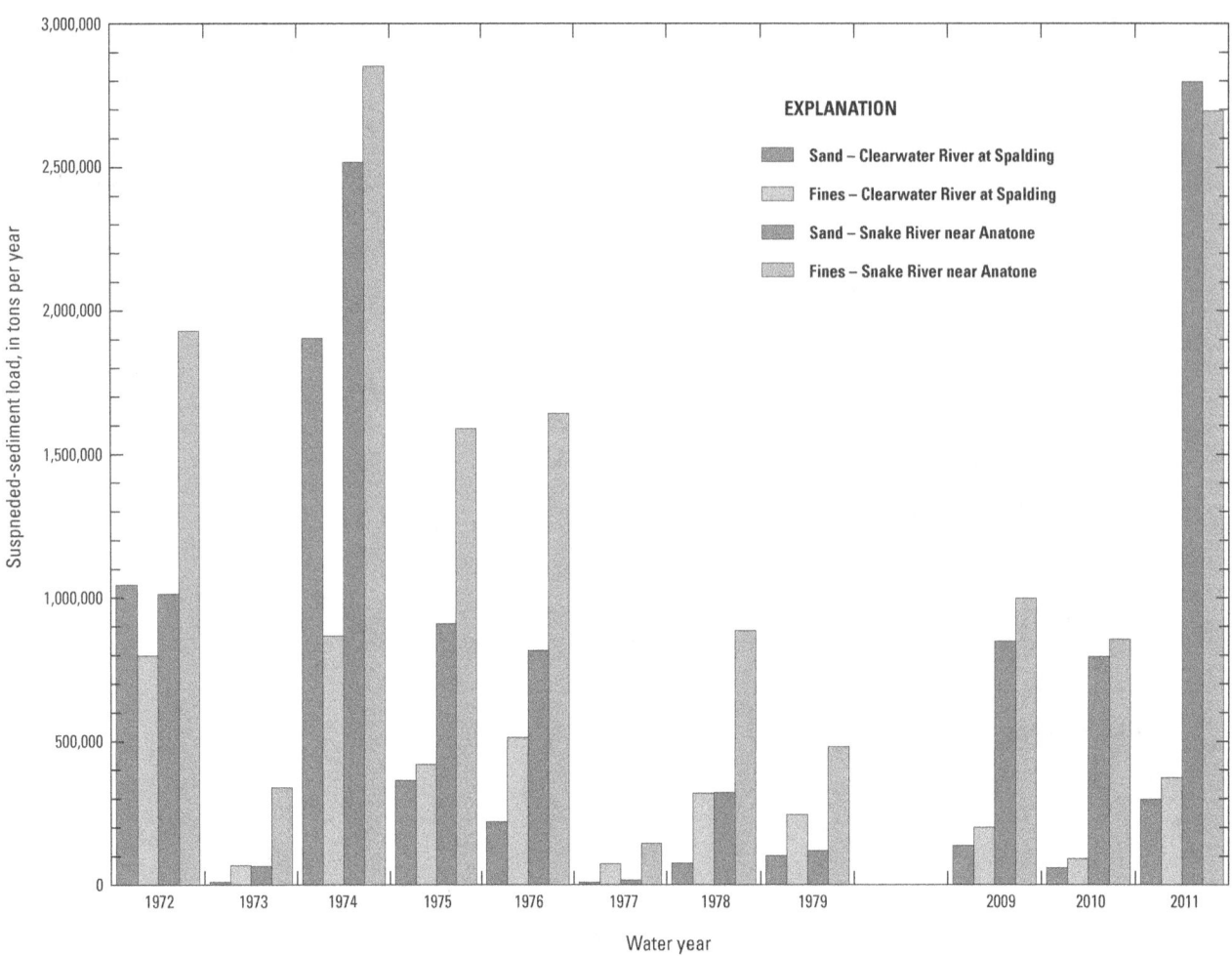

Figure 13.   Estimated annual suspended sand and fine loads, Snake River near Anatone, Washington, and Clearwater River at Spalding, Idaho, water years 1972–79 and 2009–11.

However, the increase in the suspended-sand load noted in the Snake River near Anatone probably is attributable to the Salmon River. A century of fire suppression and other forest-management practices resulted in an increase in the number and severity of forest fires in central Idaho during the last quarter of the 20th century (Burton, 2005) and the first decade of the 21st century. The effect of wild fires on sediment mobility can be particularly dramatic in the Salmon River Basin and other areas of central Idaho where disturbance of steep drainage basins with highly erosive soils can mobilize large quantities of sand and gravel to streams (King and others, 2004). Although it is beyond the scope of this study, it would be insightful to obtain fluvial sediment data from select locations within the Salmon River Basin to identify critical subbasins and to quantify the magnitude of sediment delivery to discrete reaches of the Salmon River, the lower Snake River, and ultimately to Lower Granite Reservoir.

Suspended-Sediment Surrogates

Surrogate models were developed using suspended sediment and acoustic backscatter data collected during water year 2010, and data from water year 2011 were used to validate the models. The best surrogate models based on the highest R^2 were developed using the 1.5MHz ADVM at the streamgage on the Snake River near Anatone and the 3MHz ADVM at the streamgage on the Clearwater River at Spalding (table 5; fig. 14). Separate models were developed for each streamgage to estimate overall SSC as well as concentrations of sand and fines (table 5). SSC results from water year 2011 matched the acoustic surrogate models, on average, -8.3 percent at the Snake River station and +9.8 percent at the Clearwater River station. Major deviations from the models were not evident in water year 2011 except for one sample with a low sediment concentration at the Clearwater River gaging station (fig. 14).

Table 5. Summary of acoustic surrogate and LOADEST models used to evaluate suspended-sediment concentrations at the Snake River near Anatone, Washington, and Clearwater River at Spalding, Idaho.

[Locations of stations are shown in figure 2. **Abbreviations:** R^2, coefficient of determination; RPD, relative percent difference between sample and model results; BCF, Duan's bias correction factor; MHz, megahertz; ABScorr, acoustic backscatter corrected for beam spreading and attenuation by water and sediment (dB, decibel); ln, natural log; SSC, suspended-sediment concentration (milligrams per liter); Q, streamflow (cubic feet per second); T, is the centered decimal time in years from the beginning of the calibration period]

Gaging station name	Model	Number of samples used for regression; validation	Model	R^2	Average RPD (percent)	Standard error	BCF
Snake River near Anatone	1.5MHz_ABS$_{corr}$	22; 9	SSC = $10^{[(0.0756 \times 1.5MHz\_ABScorr)-4.676]} \times 1.048$	0.92	+10	1.39	1.048
			Sand concentration = $10^{[(0.105 \times 1.5MHz\_ABScorr)-7.636]} \times 1.129$	0.89	+24	1.73	1.129
			Fines concentration = $10^{[(0.0615 \times 1.5MHz\_ABScorr)-3.730]} \times 1.084$	0.81	+19	1.54	1.084
	LOADEST	38	ln SSC = $3.96 + 2.09(\ln Q) + 0.08(\ln Q^2) - 0.54[\sin(2\pi T)] + 0.16[\cos(2\pi T)]$	0.90			
			ln sand conc = $2.97 + 2.99(\ln Q) - 0.46(\ln Q^2) - 0.49[\sin(2\pi T)] - 0.06[\cos(2\pi T)]$	0.95			
			ln fines conc = $3.38 + 1.72(\ln Q) + 0.22(\ln Q^2) - 0.56[\sin(2\pi T)] + 0.34[\cos(2\pi T)]$	0.80			
Clearwater River at Spalding	3MHz_ABS$_{corr}$	30; 11	SSC = $10^{[(0.0557*3MHz\_ABScorr)-2.431]} \times 1.040$	0.93	+8.6	1.34	1.040
			Sand concentration = $10^{[(0.0743 \times 3MHz\_ABScorr)-4.147]} \times 1.146$	0.87	+34	1.71	1.146
			Fines concentration = $10^{[(0.0461 \times 3MHz\_ABScorr)-2.030]} \times 1.097$	0.78	+19	1.58	1.097
	LOADEST	38	ln SSC = $1.79 + 1.05(\ln Q) + 0.59(\ln Q^2) + 0.46[\sin(2\pi T)] - 0.48[\cos(2\pi T)]$	0.74			
			ln sand conc = $0.05 + 1.53(\ln Q) + 0.69(\ln Q^2) + 0.24[\sin(2\pi T)] - 0.85[\cos(2\pi T)]$	0.84			
			ln fines conc = $1.53 + 0.85(\ln Q) + 0.52(\ln Q^2) + 0.65[\sin(2\pi T)] - 0.36[\cos(2\pi T)]$	0.62			

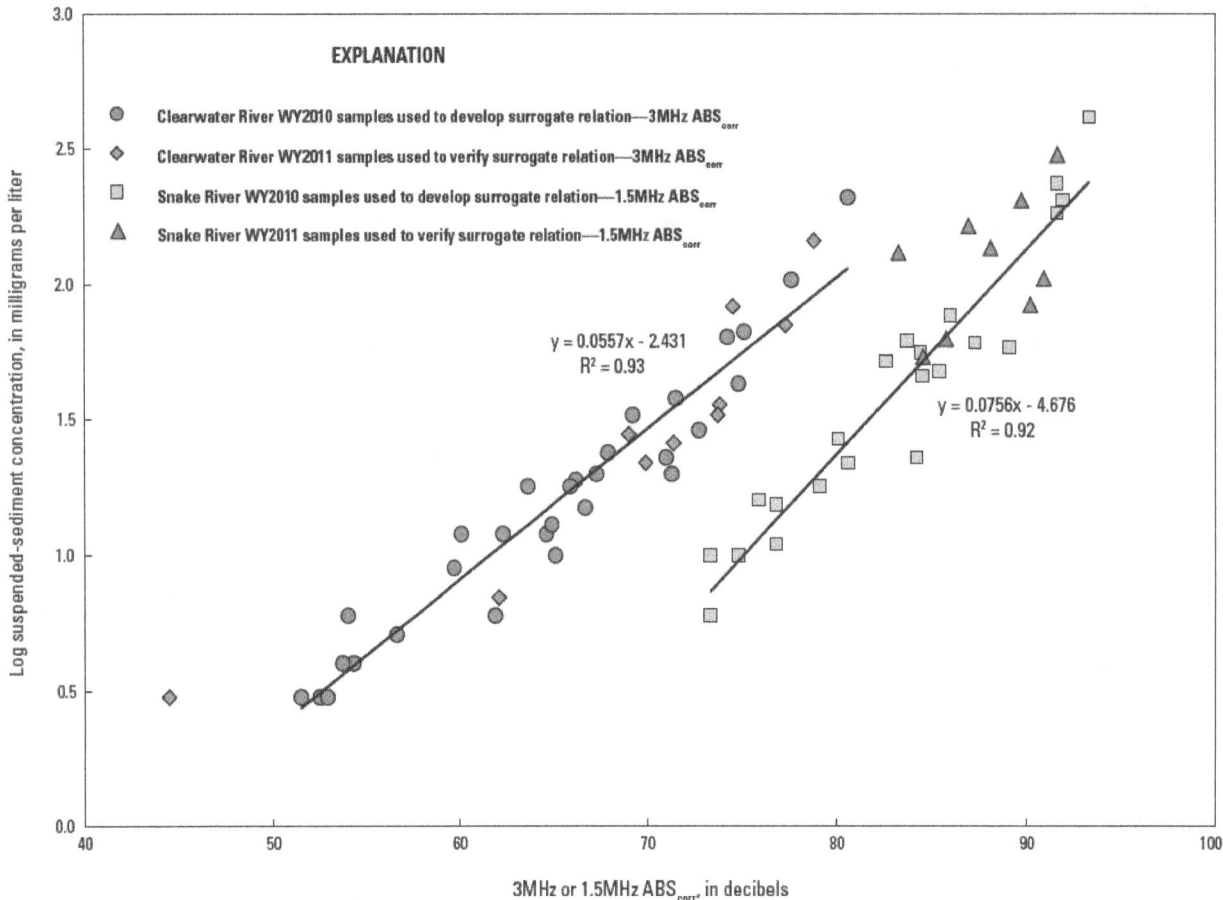

Figure 14.   Best-fit surrogate models for the Snake River near Anatone, Washington, using the 1.5MHz acoustic Doppler velocity meter (ADVM) and the Clearwater River at Spalding, Idaho, using the 3MHz ADVM, water years 2010–11.

Suspended-sediment loads computed using the acoustic surrogate models for the Snake and Clearwater Rivers were compared to the loads calculated using LOADEST. The acoustic surrogate models were developed for concentration only, and loads were calculated by multiplying concentration estimates by daily streamflow as measured at each streamgage. Performance of the models was compared by first evaluating the coefficient of determination (R^2) of the acoustic surrogate and LOADEST concentration (not load) models. The acoustic surrogate models had a better correlation than the LOADEST concentration models at both stations for SSC and suspended fines and for suspended sand at the Clearwater River station (table 5). The LOADEST concentration model had a better correlation than the acoustic surrogate models for suspended sands at the Snake River station (table 5).

Load results for the Snake River using LOADEST and acoustic surrogate models were markedly different. The largest difference occurred in water year 2011, when the suspended-sediment load calculated using the LOADEST model was more than three times higher than the load calculated using the acoustic surrogate model (table 6). The total monthly sediment loads were substantially higher using the LOADEST model as compared to the acoustic surrogate model, particularly during May and June (fig. 15A). Generally, SSC at the Snake River station were underestimated by the acoustic surrogate model when concentrations were high (greater than 200 mg/L) and had a high percentage of sand. The unregulated Salmon River contributed an average 59 percent of the streamflow passing the Snake River at Anatone streamgage during these sampling events, but there was no clear relation between the percentage

Table 6. Comparison of suspended-sediment loads estimated using acoustic backscatter and LOADEST models in the Snake River near Anatone, Washington, and Clearwater River at Spalding, Idaho.

[Locations of stations are shown in figure 2. **Total suspended-sediment load** may not equal total sand load plus total fines load due to errors in individual models. **Abbreviations:** MHz, megahertz; ABS$_{corr}$, acoustic backscatter corrected for beam spreading and attenuation by water and sediment (dB, decibel); Q, streamflow]

Gaging station name	Water year	Model	Total suspended-sediment load (tons)	Total suspended-sand load (tons)	Total suspended-fines load (tons)
Snake River near Anatone	2009	1.5MHz_ABS$_{corr}$ [1]	1,460,000	709,000	789,000
		LOADEST (Q, Q^2, seasonality)	1,820,000	848,000	998,000
	2010	1.5MHz_ABS$_{corr}$	1,120,000	589,000	606,000
		LOADEST (Q, Q^2, seasonality)	1,650,000	795,000	855,000
	2011	1.5MHz_ABS$_{corr}$	1,740,000	701,000	1,030,000
		LOADEST (Q, Q^2, seasonality)	5,450,000	2,800,000	2,700,000
	Overall	1.5MHz_ABS$_{corr}$	4,320,000	2,000,000	2,430,000
		LOADEST (Q, Q^2, seasonality)	8,920,000	4,440,000	4,550,000
Clearwater River at Spalding	2009	3MHz_ABS$_{corr}$	315,000	138,000	180,000
		LOADEST (Q, Q^2, seasonality)	335,000	137,000	199,000
	2010	3MHz_ABS$_{corr}$	169,000	73,900	98,200
		LOADEST (Q, Q^2, seasonality)	152,000	59,000	91,200
	2011	3MHz_ABS$_{corr}$	645,000	329,000	342,000
		LOADEST (Q, Q^2, seasonality)	662,000	297,000	372,000
	Overall	3MHz_ABScorr	1,130,000	541,000	620,000
		LOADEST (Q, Q^2, seasonality)	1,150,000	493,000	662,000

[1]Loads presented for the 1.5MHz_ABS$_{corr}$ in water year 2009 are for April–September only.

of unregulated streamflow passing the station and percent error in the estimated SSCs. There also was no clear relation between percent organic matter and percent error in the estimations. Some of the error in the acoustic surrogate model at high sand concentrations may be due to the placement of the ADVM in the water column. If the sand particles were not well-mixed in the water column during specific events or were transported in a layer close to the streambed, the ADVM would not have detected those particles because it is mounted approximately mid-depth in the water column. This scenario is likely, particularly in water year 2011, because of the large amount of sand transport measured in the Salmon River and contributed to the Snake River upstream of the surrogate monitoring station at Anatone.

Further examination of estimated SSCs in water year 2011 shows that the LOADEST model produces much higher estimated concentrations than the sediment surrogate model on the descending limb of the hydrograph (fig. 16). The acoustic surrogate model should be a more direct measurement of sediment concentrations as compared to the LOADEST model (which is based on flow, time, and seasonality terms), at least during storm events. The pattern over the ascending and descending limbs of the hydrograph show that the LOADEST model is affected by hysteresis, but perhaps less so than a traditional transport curve based on a single streamflow term (Wood and Teasdale, 2013). Generally, the acoustic surrogate model matched more closely with the measured SSCs in samples collected at the Snake River near Anatone than the LOADEST model, except during high streamflow in May and June 2011 (fig. 16). Neither the acoustic or the LOADEST model performed well during this period, especially during June when the sample results were between the concentrations estimated by the two models.

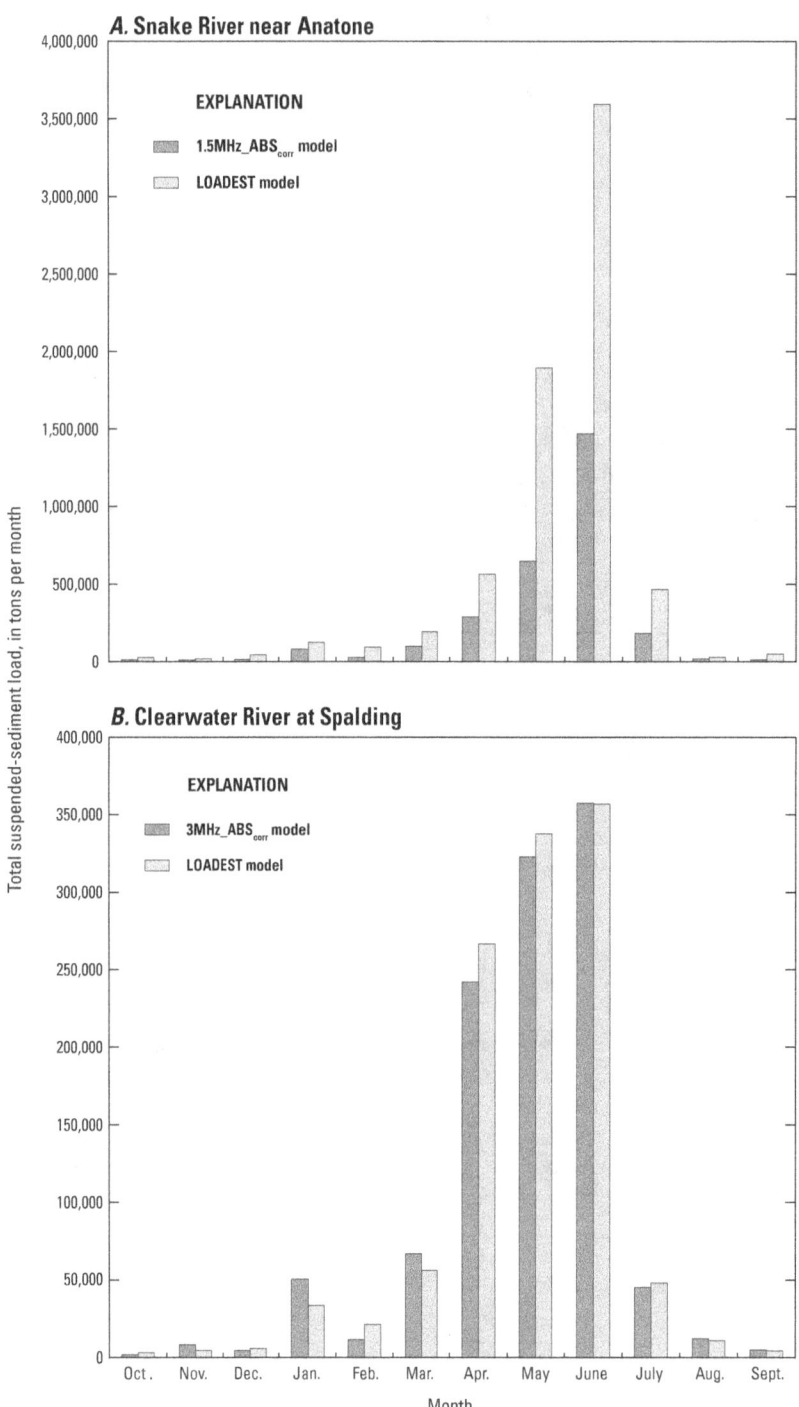

Figure 15.   Comparison of total monthly sediment load estimated using acoustic backscatter and LOADEST models in the (*A*) Snake River near Anatone, Washington, and (*B*) Clearwater River at Spalding, Idaho, water years 2009–11.

NOTE: Each month represents the total load that occurred during that month within the study period. For example, the total load for May in the Clearwater River is the sum of loads in May 2009, May 2010, and May 2011.

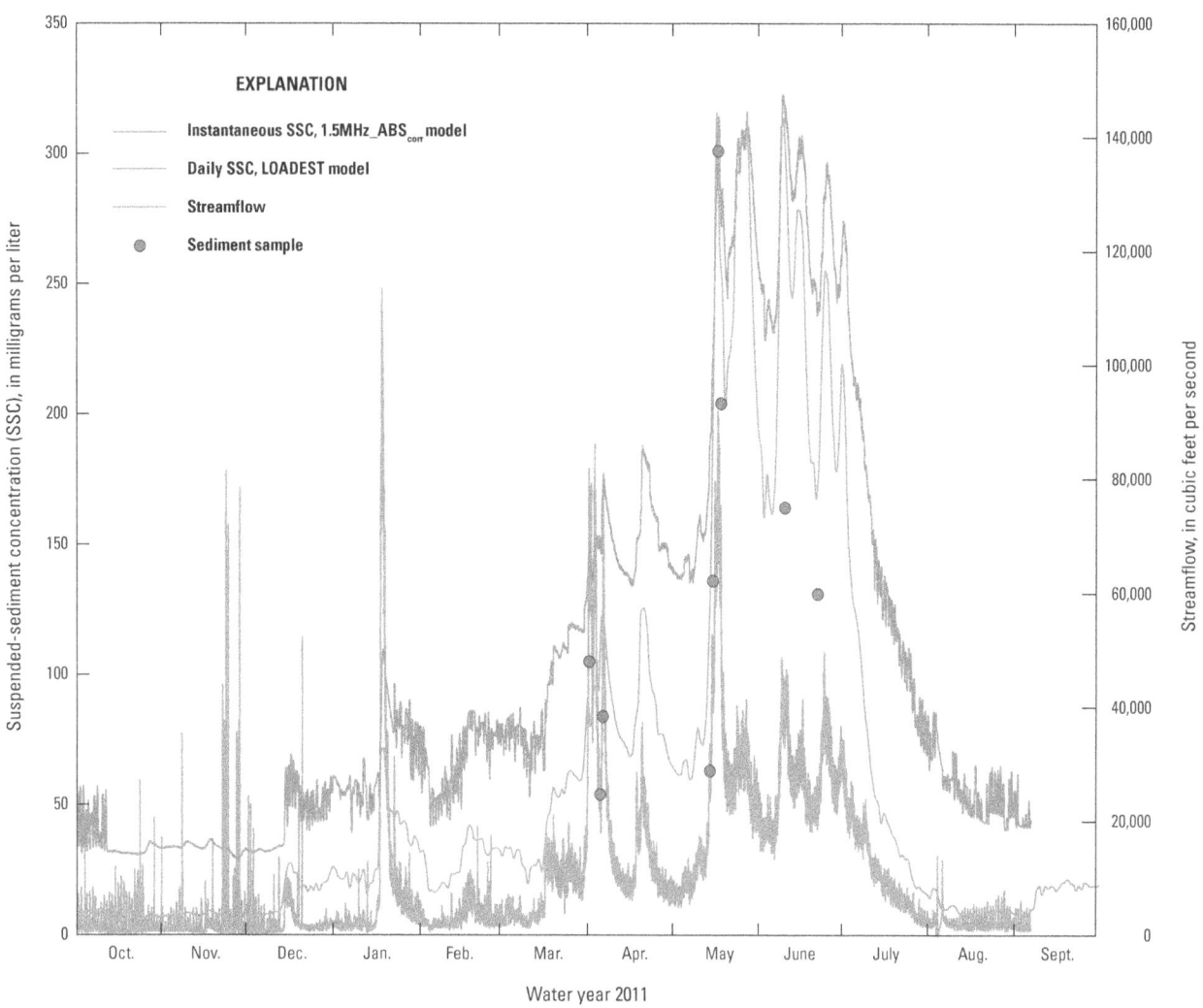

Figure 16. Comparison of suspended-sediment concentrations estimated using acoustic backscatter and a LOADEST model, Snake River near Anatone, Washington, water year 2011.

Model simulation results for load compared well for the Clearwater River station. Suspended-sediment loads at the Clearwater River computed by the acoustic surrogate and LOADEST models matched within 3–11 percent on an annual basis and within 2 percent over the entire study period (table 6). The percent differences in the Clearwater River were slightly higher for total suspended-sand load, from 1–22 percent on an annual basis. Overall, however, the model results compared well. The acoustic surrogate model, in general, produced sand load estimates using the 1.5 MHz ADVM that were slightly higher than the estimates generated using the LOADEST model (9 percent higher for the entire study period). Suspended fines loads, which were estimated using the higher frequency 3.0 MHz ADVM, compared within 7–10 percent on an annual basis between the two

models, and within 7 percent for the entire study period. The monthly average suspended-sediment loads for 2010–11 for the Clearwater River also compared well between the acoustic surrogate and LOADEST models (fig. 15B).

For storm events, the acoustic surrogate models provide a more direct measure of sediment concentrations as compared to LOADEST or traditional transport curves that are based only on streamflow. Thus, the acoustic surrogates probably are a more accurate method for estimating suspended-sediment loads over a short period or when real-time response is needed. Additionally, LOADEST models cannot be used to provide real-time, instantaneous load estimates because of post-processing requirements. Differences in load estimates between the acoustic surrogate and LOADEST models were further examined by summing estimated loads over the ascending and descending limbs of the hydrograph for several well-defined hydrologic events during the study period: seven events for the Snake River near Anatone and eight events for the Clearwater River at Spalding (table 7). Loads calculated using the acoustic surrogate model were higher on the ascending limb (negative percent difference) and lower on the descending limb (positive percent difference) as compared to loads calculated using the LOADEST model in all cases except for one event at the Snake River station (ascending limb) and two events at the Clearwater River station (one on the ascending limb, one on the descending limb). For all events combined, loads calculated using the acoustic surrogate model were 11–14 percent higher on the ascending limb and 24–65 percent lower on the descending limb as compared to loads calculated using the LOADEST model. The highest calculated and measured SSC typically were higher on the ascending limb (especially for acoustics), but total calculated loads often were higher on the descending limb than the ascending limb due to a prolonged recession in the streamflow.

During the June 2010 spring snowmelt runoff, multiple suspended-sediment samples were collected and compared to the acoustic surrogate results to validate the relative performance as compared to LOADEST models. In the Snake and Salmon River Basins, a large pulse of sediment-laden water was delivered from the Salmon River to the Snake River during this snowmelt runoff event. This sediment pulse was captured in the acoustic surrogate model as a spike in the acoustic backscatter on June 3 (fig. 17A), but not by the LOADEST model, which is primarily based on changes in streamflow. Samples were not collected on June 3, 2010 to verify either model. The acoustic surrogate model showed a rapid decrease in SSC directly following the peak of the hydrograph on June 3, whereas the LOADEST model showed elevated SSC until well after the peak. The sample with high SSC collected on June 6 does not match estimates from either model, but the samples collected on June 4 and June 14 match well with the estimated SSC from the acoustic surrogate model. The acoustic surrogate model estimated sand concentrations close to all three of the sand concentrations in the samples collected from the Snake River (fig. 17B). The largest difference between model estimates and the June 6 simulation results was for the fines concentration (fig. 17C), which was better represented by the LOADEST model. During the April 2010 hydrologic event on the Clearwater River (fig. 18), the sample SSC results match the acoustic surrogate model estimates closely, whereas the LOADEST model underestimates most concentrations and loads throughout the event (fig. 18, table 7).

Table 7. Comparison of suspended-sediment loads during selected hydrologic events calculated using acoustic backscatter and a LOADEST model in the Snake River near Anatone, Washington, and Clearwater River at Spalding, Idaho.

[Locations of stations are shown in figure 2. **ABS$_{corr}$**: Acoustic backscatter corrected for beam spreading and attenuation by water and sediment (dB, decibel). **LOADEST:** Load Estimator. Percentages in parentheses indicate negative difference. **Abbreviation:** NA, not applicable]

Gaging station name	Event No.	Dates of event	Suspended-sediment load over ascending limb of hydrograph (tons)			Suspended-sediment load over descending limb of hydrograph (tons)		
			ABScorr	LOADEST	Percent difference	ABS$_{corr}$	LOADEST	Percent difference
Snake River near Anatone	NA	NA–All seven events combined	729,000	632,000	(14)	862,000	1,700,000	65
	1	April 18–May 3, 2009	122,000	85,900	(35)	89,600	115,000	25
	2	May 5–11, 2009	48,600	37,400	(26)	24,400	37,800	43
	3	May 18–24, 2009	98,400	86,700	(13)	106,000	159,000	40
	4	April 20–27, 2010	15,700	13,000	(19)	12,100	13,100	8
	5	May 16–30, 2010	33,000	26,100	(23)	48,400	59,900	21
	6	June 2–19, 2010	267,000	214,000	(22)	481,000	953,000	66
	7	May 11–19, 2011	144,000	169,000	16	100,000	357,000	112
Clearwater River at Spalding	NA	NA–All eight events combined	274,000	245,000	(11)	225,000	287,000	24
	1	May 13–24, 2008	59,400	94,500	46	13,000	22,200	52
	2	April 21–May 1, 2009	13,000	10,700	(19)	17,700	32,600	59
	3	April 16–27, 2010	5,740	2,990	(63)	3,700	3,260	(13)
	4	May 14–26, 2010	14,700	6,880	(72)	7,370	10,700	37
	5	June 2–July 5, 2010	70,600	55,600	(24)	34,900	37,300	7
	6	January 15–February 2, 2011	17,600	5,520	(104)	17,800	18,600	4
	7	March 29–April 15, 2011	55,500	45,300	(20)	93,900	113,000	18
	8	May 11–20, 2011	37,400	23,400	(46)	36,200	49,600	31

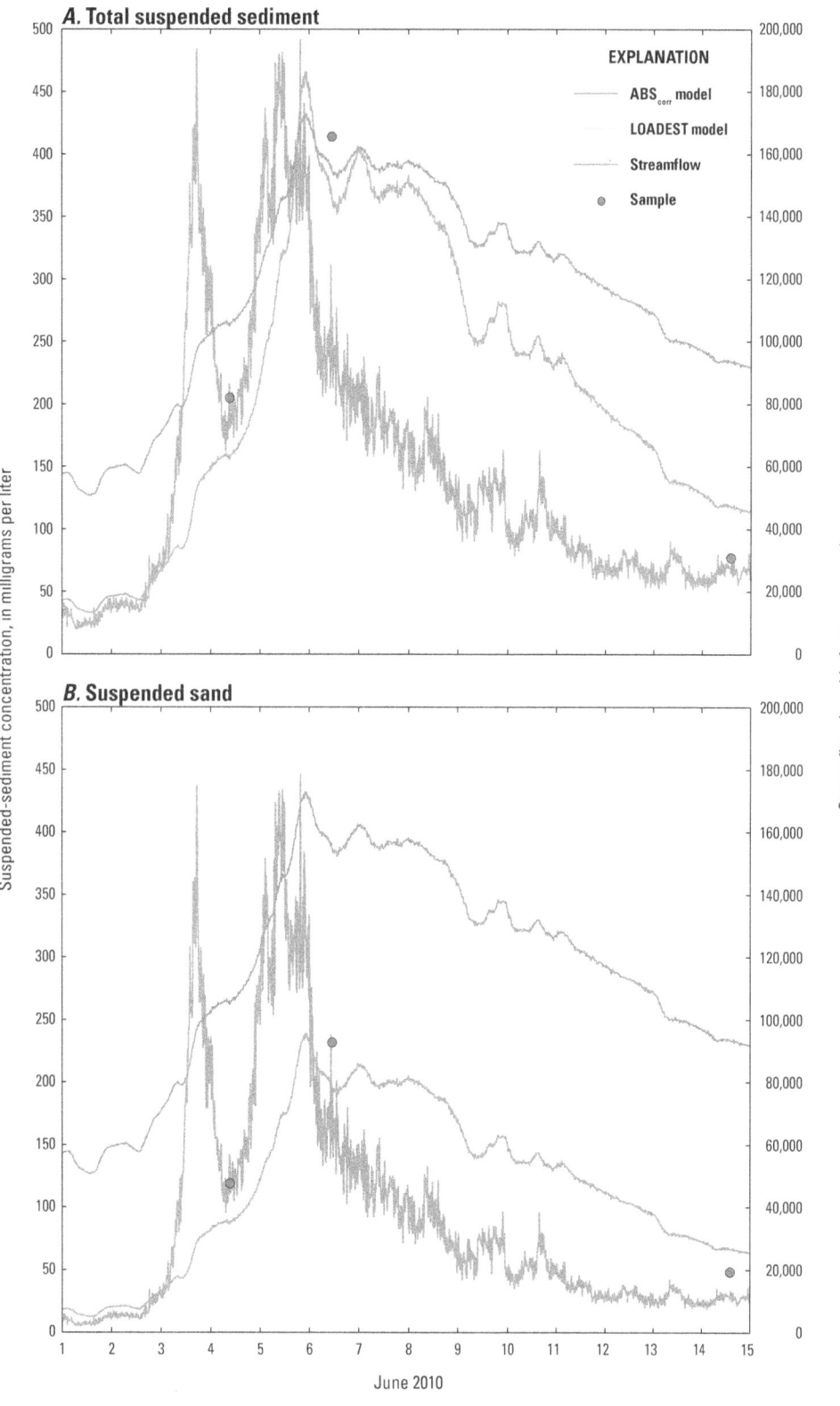

Figure 17. Comparison of (*A*) total suspended sediment, (*B*) suspended sand, and (*C*) suspended fine concentrations estimated using acoustic backscatter and a LOADEST model during spring snowmelt runoff at Snake River near Anatone, Washington, June 2010.

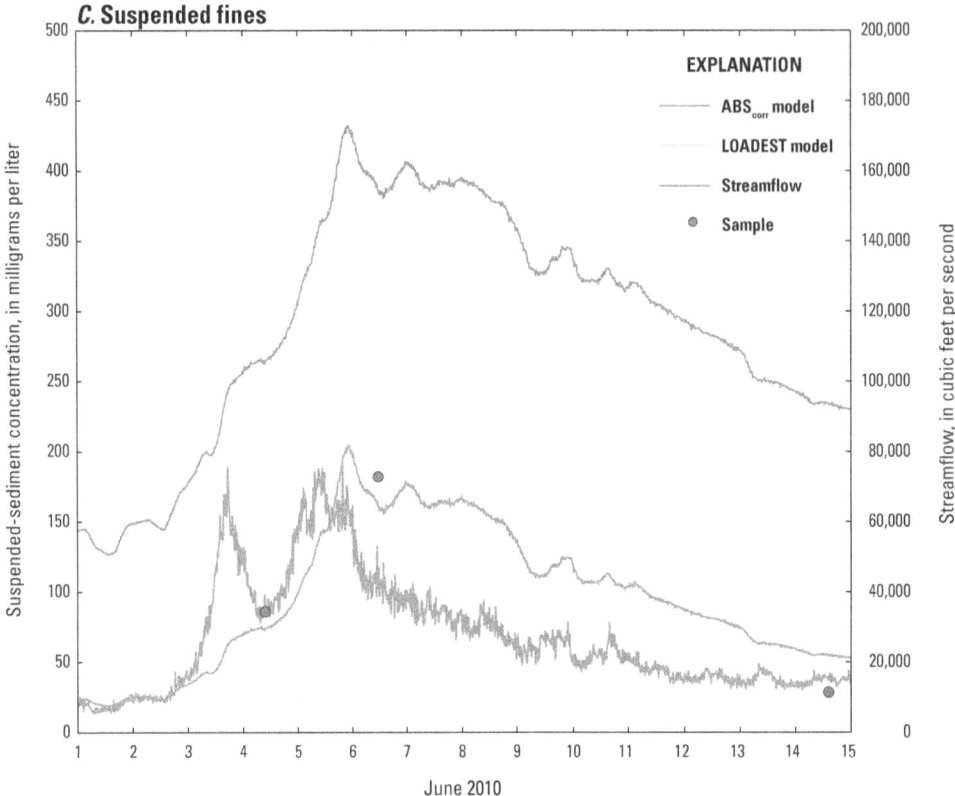

Figure 17.—Continued

Overall, in the Snake River near Anatone and the Clearwater River at Spalding, acoustic backscatter appeared to better estimate the SSC and suspended-sediment load than the LOADEST model over short periods. This is primarily because acoustic backscatter (1) is not affected by hysteresis; (2) provides a more direct, in-situ measurement of suspended sediment; and (3) is better able to represent sediment sources from a combination of regulated and unregulated sources, which can be poorly represented by a model based primarily on changes in streamflow. However, on a monthly and (or) annual basis, the acoustic surrogate and LOADEST models produced similar sediment load estimates for the Clearwater River at Spalding. Which model produced better estimates of monthly and annual sediment loads for the Snake River near Anatone is inconclusive. Much of the difference between the model simulation results is represented on the falling limb of the hydrograph and at mid-range streamflows during which few samples were collected. However, during high streamflow

events such as the snowmelt runoff in the Snake River during 2011, acoustic surrogate tools may be unable to capture the contribution of suspended sand moving near the bottom of the water column and thus, may underestimate the total load of suspended sediment.

It is useful to show the acoustic surrogate model simulation results for comparison with LOADEST models as a demonstration that the most desirable method for estimating sediment load depends on the period of interest, the type of sediment being transported, and the degree of mixing and (or) stratification of the sediment in the stream. On an annual basis, both the LOADEST and surrogate models appear to provide a good estimate of suspended-sediment loads in the Snake and Clearwater Rivers. However, because surrogate methods for estimating loads were not available in all the tributary basins evaluated during this study, the LOADEST simulation results were used for the basin-wide sediment budgets to provide a consistent method for comparison.

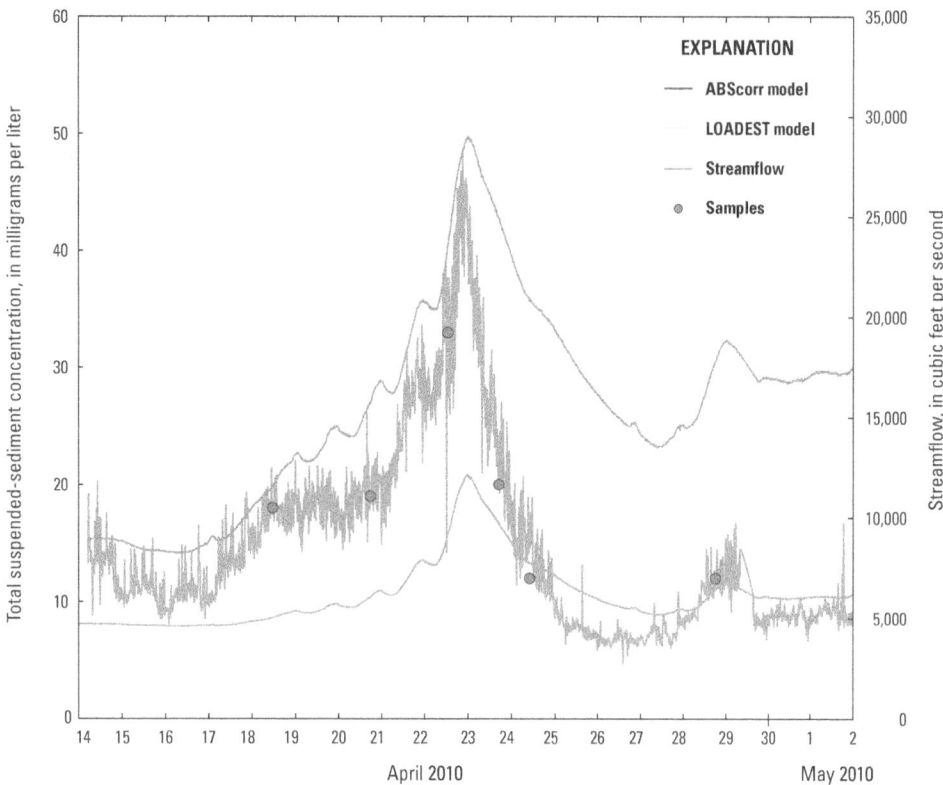

Figure 18. Comparison of total suspended-sediment concentrations estimated using acoustic backscatter and a LOADEST model during a hydrologic event at Clearwater River at Spalding, Idaho, April and May 2010.

Bedload

Bedload was highly variable throughout the Snake and Clearwater River Basins and at each station where samples were collected (table 8). Measured bedload ranged from zero during base flow conditions at all stations to a maximum of 4,140 ton/d in the Salmon River at White Bird at a streamflow of 70,800 ft³/s (table 8). Stations located in upstream areas of a drainage basin generally had a larger percentage of the total sediment load as bedload (fig. 19) and a stronger correlation (higher R^2) between streamflow and bedload discharge as compared to stations located downstream (table 9). High stream gradients and high stream velocities at upstream stations probably account for the higher percentage of bedload. Best-fit regression equations and coefficients of determination (R^2) for the stations where bedload was collected are presented in table 9. Bedload transport curves are shown in figure 20. At most of the stations with measureable bedload, the particle size distribution (table 10) typically was unimodal at lower

streamflows, with sand being the dominant particle size. At higher streamflows and during peak bedload discharge, the particle size was typically bimodal, with sand and coarse gravel composing the predominant particle size.

In the Snake River Basin, the measured bedload at the Salmon River at White Bird ranged from 25.9 to 4,140 ton/d (table 8). However, only five bedload samples were collected from the Salmon River during this study, and the correlation between streamflow and bedload was relatively poor ($R^2 = 0.46$; table 9). Three of the five samples collected from the Salmon River were unimodal, with 99 percent of the bedload composed of sand. The two other bedload samples from the Salmon River were bimodal and consisted of 35–45 percent gravel (2–64 mm diameter) and 55–65 percent sand (0.063–2 mm diameter) (table 10). Based on the best-fit regression equation (table 9), the total bedload discharge in the Salmon River was about 66,000 tons during water years 2009–11; about 1.3 percent of the total sediment load transported in the Salmon River at White Bird (fig. 19).

Table 8. Bedload data from sampling stations in the lower Snake and Clearwater River Basins, Washington and Idaho, water years 2008–11.

[Locations of stations are shown in figure 2. **Sampler type:** HS, Helley Smith. **Abbreviations:** ft, foot; ft³/s, cubic foot per second; ton/d, ton per day]

Date	Mean sample time	Channel width (ft)	Streamflow, instantaneous (ft³/s)	Number of sampling points	Rest time on bottom (seconds)	Sampler type	Bedload discharge (ton/d)
Salmon River at White Bird, Idaho (13317000)							
04-24-09	1030	370	32,400	10	60	HS	188
05-20-09	1332	460	70,800	10	60	HS	4,140
06-03-09	1535	445	60,400	10	60	HS	189
06-11-09	0821	400	41,000	10	60	HS	25.9
05-15-11	1245	430	55,300	7	60	HS	545
Snake River near Anatone, Washington (13334300)							
05-06-08	1600	560	49,800	20	60	HS	110
06-10-08	1145	597	73,200	10	60	HS	271
07-08-08	0920	550	40,900	10	60	HS	14.2
04-16-09	0915	570	46,600	20	60	HS	16
04-23-09	1230	590	83,000	10	60	HS	214
05-20-09	1330	630	105,000	20	60	HS	89.8
06-16-09	1750	600	72,100	20	60	HS	165
05-20-10	0916	580	62,500	8	60	HS	176
06-04-10	1425	617	105,000	20	60	HS	195
06-14-10	1518	622	93,700	20	60	HS	170
04-04-11	1519	570	68,000	20	60	HS	406
04-05-11	1310	570	79,900	20	60	HS	317
05-14-11	1555	589	102,000	14	60	HS	20.4
05-16-11	1335	608	142,000	14	60	HS	346
05-17-11	1550	608	125,000	13	60	HS	183
06-21-11	1248	590	110,000	15	60	HS	491
Selway River near Lowell, Idaho (13336500)							
04-22-09	1045	270	14,000	20	60	HS	17.5
05-18-09	1500	300	18,500	20	60	HS	234
05-19-09	1040	315	27,300	20	60	HS	463
05-21-09	1123	310	20,300	20	60	HS	75.2
06-09-09	1231	303	13,800	20	60	HS	72.1
05-19-10	1110	306	16,400	20	60	HS	24.9
06-04-10	1350	308	21,800	20	60	HS	440
06-15-10	1318	300	14,200	20	60	HS	29.4
05-17-11	1025	305	19,400	20	60	HS	151
05-24-11	1730	308	22,000	20	60	HS	92
06-10-11	0816	305	20,600	20	30	HS	86.1
Lochsa River near Lowell, Idaho (13337000)							
05-18-11	0815	276	12,600	20	60	HS	132
06-07-11	0905	275	21,200	20	60	HS	329
06-10-11	1036	280	14,800	20	60	HS	26.0

Table 8. Bedload data from sampling stations in the lower Snake and Clearwater River Basins, Washington and Idaho, water years 2008–11.—Continued

[Locations of stations are shown in <u>figure 2</u>. **Sampler type:** HS, Helley Smith. **Abbreviations:** ft, foot; ft³/s, cubic foot per second; ton/d, ton per day]

Date	Mean sample time	Channel width (ft)	Streamflow, instantaneous (ft³/s)	Number of sampling points	Rest time on bottom (seconds)	Sampler type	Bedload discharge (ton/d)
\multicolumn{8}{c}{Middle Fork Clearwater River at Kooskia, Idaho (13337120)}								
04-22-09	1500	465	25,600	20	60	HS	23.6
05-18-09	1523	455	34,000	20	60	HS	336
05-19-09	1530	490	46,000	14	60	HS	222
05-21-09	1653	460	35,000	20	60	HS	258
06-09-09	1600	460	24,000	20	60	HS	13.9
05-21-10	1607	457	22,700	20	60	HS	13.5
06-04-10	1500	495	44,100	20	60	HS	163
06-16-10	1011	455	23,400	20	60	HS	33.7
05-17-11	1600	470	37,000	20	60	HS	163
05-26-11	0830	500	46,700	20	60	HS	146
06-08-11	1044	508	57,000	20	60	HS	592
\multicolumn{8}{c}{South Fork Clearwater River near Harpster, Idaho (13338100)}								
04-08-09	1021	72	1,410	20	60	HS	0.6
04-14-09	1252	66	2,340	20	60	HS	10.3
04-22-09	0937	80	5,540	20	60	HS	16.7
05-07-09	0930	82	6,340	6	60	HS	79
05-13-09	1230	76	4,000	20	60	HS	5.8
05-21-09	1145	80	6,070	12	60	HS	103
06-10-09	1414	63	3,190	17	60	HS	19.2
04-22-10	1507	71	2,450	20	60	HS	11.4
05-19-10	1525	72	2,830	20	60	HS	10.8
05-20-10	1216	70	2,820	20	60	HS	12.3
06-06-10	0840	80	4,650	20	60	HS	21.4
06-14-10	1328	76	3,270	20	60	HS	18.2
05-10-11	1057	75	3,900	20	60	HS	12
06-11-11	1000	82	6,750	18	60	HS	29
06-28-11	0903	71	3,470	17	60	Elwha	12.9
\multicolumn{8}{c}{South Fork Clearwater River at Stites, Idaho (13338500)}								
04-14-09	1646	150	2,930	20	60	HS	4.4
04-22-09	1402	154	5,310	20	60	HS	30.6
05-07-09	1430	144	6,720	20	60	HS	190
05-13-09	1630	154	4,020	20	60	HS	32.6
05-21-09	1430	160	5,470	20	60	HS	155
06-10-09	0833	153	3,260	20	60	HS	18.8
04-22-10	1925	151	2,450	20	60	HS	10.2
05-19-10	1822	150	2,640	20	60	HS	10.1
05-20-10	0847	150	3,480	20	60	HS	18.4
06-05-10	1425	158	6,110	20	60	HS	208
06-15-10	1803	150	3,090	20	60	HS	31.2
05-10-11	1545	155	4,220	20	60	HS	12.8
06-10-11	1555	161	7,490	20	60	HS	603
06-28-11	1210	147	3,480	20	60	Elwha	31.2

Table 8.   Bedload data from sampling stations in the lower Snake and Clearwater River Basins, Washington and Idaho, water years 2008–11.—Continued

[Locations of stations are shown in figure 2. **Sampler type:** HS, Helley Smith. **Abbreviations:** ft, foot; ft^3/s, cubic foot per second; ton/d, ton per day]

Date	Mean sample time	Channel width (ft)	Streamflow, instantaneous (ft^3/s)	Number of sampling points	Rest time on bottom (seconds)	Sampler type	Bedload discharge (ton/d)
\multicolumn{8}{c}{Clearwater River at Orofino, Idaho (13340000)}
04-23-09	0953	360	38,500	20	60	HS	2.64
05-19-09	1612	375	57,700	20	60	HS	44.4
06-02-09	1809	372	46,000	20	60	HS	378
06-04-10	0948	375	49,900	20	60	HS	381
06-10-10	0917	373	45,800	20	60	HS	217
06-06-11	1428	380	49,300	20	60	HS	754
06-28-11	1518	365	37,400	20	60	Elwha	23.9
\multicolumn{8}{c}{Clearwater River at Spalding, Idaho (13342500)}
05-05-08	1500	440	29,300	20	60	HS	11.9
05-19-08	1430	460	78,900	9	30	HS	841
05-28-08	1015	450	46,700	20	60	HS	149
06-11-08	1345	440	49,700	20	60	HS	155
07-07-08	1345	440	28,000	20	60	HS	6.0
04-15-09	1145	440	37,700	20	60	HS	15.2
04-23-09	1545	450	56,300	20	60	HS	43.3
05-19-09	0955	450	60,600	20	60	HS	12
06-03-09	0949	450	51,100	20	60	HS	383
05-20-10	1532	444	46,200	20	60	HS	126
06-15-11	1015	432	54,500	10	30	Elwha	21
\multicolumn{8}{c}{Palouse River at Hooper, Washington (13351000)}
03-24-09	1133	142	4,240	20	60	HS	36.8
01-17-11	1121	145	5,670	20	60	HS	5.1
01-18-11	1035	148	7,150	20	60	HS	2.5

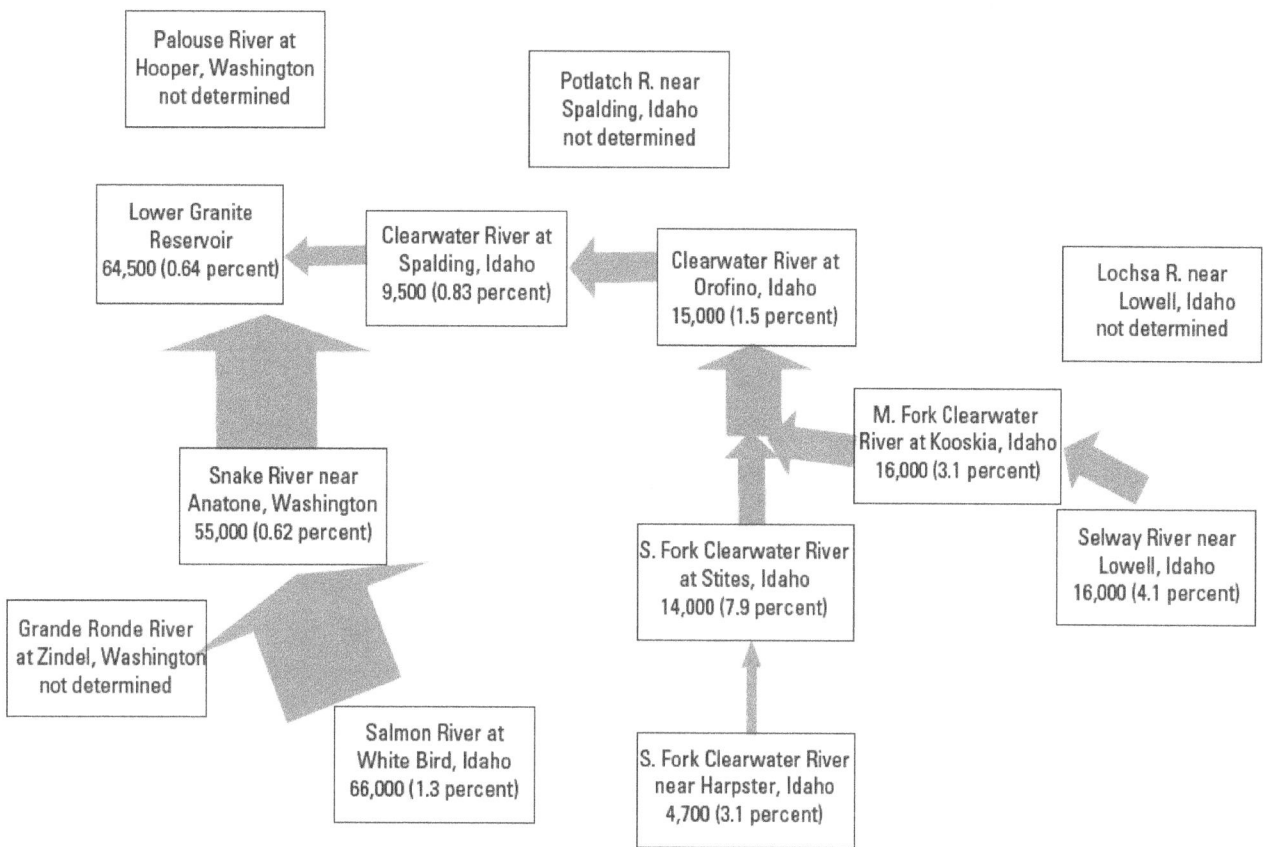

Figure 19. Estimated bedload transported in the lower Snake and Clearwater River Basins, Washington and Idaho, water years 2009–11. Values are in tons and percentage of total sediment load (suspended and bedload) transported during water years 2009–11. Width of each arrow is proportional to the estimated bedload.

Table 9. Best-fit regression equations for bedload from sampling stations in the lower Snake and Clearwater River Basins, Washington and Idaho, water years 2008–11.

[Locations of stations are shown in figure 2. Only stations with at least five samples are shown. R^2: coefficient of determination. **Abbreviations:** USGS, U.S. Geological Survey; BS, bedload discharge, in tons per day; Q, streamflow, in cubic feet per second]

USGS gaging station No.	Gaging station name	Best-fit regression equation	R^2
13317000	Salmon River at White Bird, Idaho	$\log BS = 5.033 \times 10^{-17} Q^{3.992}$	0.46
13334300	Snake River near Anatone, Washington	$\log BS = 1.842 \times 10^{-06} Q^{1.603}$	0.26
13336500	Selway River near Lowell, Idaho	$\log BS = 1.732 \times 10^{-15} Q^{3.919}$	0.61
13337120	Middle Fork Clearwater River at Kooskia, Idaho	$\log BS = 8.220 \times 10^{-10} Q^{2.644}$	0.30
13338100	South Fork Clearwater River near Harpster, Idaho	$\log BS = 2.639 \times 10^{-15} Q^{3.652}$	0.77
13338500	South Fork Clearwater River at Stites, Idaho	$\log BS = 1.081 \times 10^{-11} Q^{3.470}$	0.79
13340000	Clearwater River at Orofino, Idaho	$\log BS = 7.707 \times 10^{-33} Q^{7.312}$	0.30
13342500	Clearwater River at Spalding, Idaho	$\log BS = 6.767 \times 10^{-15} Q^{3.403}$	0.42

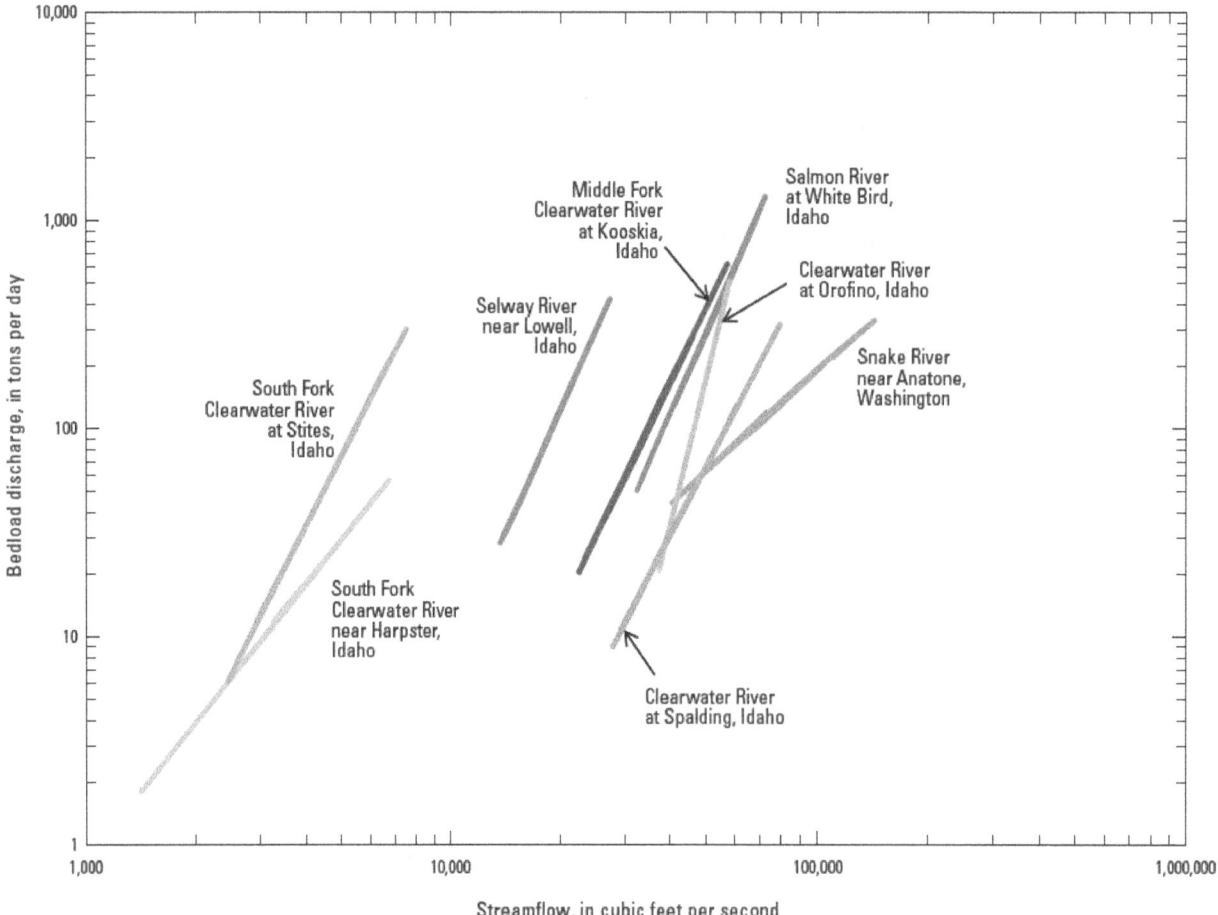

Figure 20. Bedload transport curves representing best-fit regression equations for selected stations in the lower Snake and Clearwater River Basins, Washington and Idaho.

Table 10.   Particle-size distribution for bedload samples collected from stations in the lower Snake and Clearwater River Basins, Washington and Idaho, water years 2008–11.

[Locations of stations are shown in figure 2. **Abbreviations:** <, less than; mm, millimeter]

Date	Mean sample time	< 128 mm (percent)	< 63.0 mm (percent)	< 31.5 mm (percent)	< 16.0 mm (percent)	< 8.0 mm (percent)	< 4.0 mm (percent)	< 2.0 mm (percent)	< 1.0 mm (percent)	< 0.50 mm (percent)	< 0.25 mm (percent)	< 0.125 mm (percent)	< 0.063 mm (percent)
Salmon River at White Bird, Idaho (13317000)												
04-24-09	1030	100	100	100	100	100	100	99	94	23	1	0	0
05-20-09	1332	100	96	84	67	61	57	55	52	18	2	0	0
06-03-09	1535	100	100	83	74	69	66	65	61	12	3	0	0
06-11-09	0821	100	100	100	100	100	99	98	94	7	0	0	0
05-15-11	1245	100	100	100	100	100	100	99	97	55	5	1	0
Snake River near Anatone, Washington (13334300)												
05-06-08	1600	100	100	66	54	53	53	52	50	36	1	0	0
06-10-08	1145	100	100	88	83	69	64	63	60	45	1	0	0
07-08-08	0920	100	100	100	100	100	100	99	96	73	4	0	0
04-16-09	0915	100	100	100	98	90	82	75	63	3	0	0	0
04-23-09	1230	100	81	32	28	27	26	24	21	2	0	0	0
05-20-09	1330	100	100	100	100	100	99	96	88	46	6	1	0
06-16-09	1750	100	100	90	71	56	46	40	32	6	0	0	0
05-20-10	0916	100	100	78	71	67	63	59	54	4	0	0	0
06-04-10	1425	100	100	82	74	63	59	57	52	26	3	0	0
06-14-10	1518	100	82	82	76	71	67	63	58	4	0	0	0
04-04-11	1519	100	69	34	15	10	9	8	7	1	0	0	0
04-05-11	1310	100	60	29	11	4	3	3	2	0	0	0	0
05-14-11	1555	100	100	17	17	14	13	13	12	10	1	0	0
05-16-11	1335	100	100	89	60	35	22	17	13	6	0	0	0
05-17-11	1550	100	77	34	29	19	12	8	6	0	0	0	0
06-21-11	1248	100	90	29	9	5	2	0	0	0	0	0	0
Selway River near Lowell, Idaho (13336500)												
04-22-09	1045	100	100	100	90	88	87	85	70	3	0	0	0
05-18-09	1500	100	100	100	98	98	97	96	68	3	0	0	0
05-19-09	1040	100	86	71	63	61	59	57	38	4	0	0	0
05-21-09	1123	100	83	66	48	38	34	33	26	1	0	0	0
06-09-09	1231	100	100	100	99	92	85	76	35	0	0	0	0
05-19-10	1110	100	100	100	100	100	100	98	82	7	0	0	0
06-04-10	1350	100	91	83	81	80	78	75	48	2	0	0	0
06-15-10	1318	100	100	92	89	88	86	84	53	1	0	0	0
05-17-11	1025	100	100	100	94	89	85	80	50	14	1	0	0
05-24-11	1730	100	100	86	82	80	78	75	49	14	0	0	0
06-10-11	0816	100	100	56	51	50	49	46	29	1	0	0	0
Lochsa River near Lowell, Idaho (13337000)												
05-18-11	0815	100	82	66	56	45	38	33	20	6	1	0	0
06-07-11	0905	100	82	81	79	76	74	70	45	4	0	0	0
06-10-11	1036	100	100	82	54	43	38	33	20	6	0	0	0
Middle Fork Clearwater River at Kooskia, Idaho (13337120)												
04-22-09	1500	100	100	100	100	100	99	97	79	3	0	0	0
05-18-09	1523	100	73	25	16	15	14	14	11	1	0	0	0
05-19-09	1530	100	100	57	49	48	47	45	37	15	1	0	0
05-21-09	1653	100	100	99	95	91	87	84	65	3	0	0	0
06-09-09	1600	100	100	100	100	97	92	86	59	2	0	0	0
05-21-10	1607	100	100	100	100	97	91	88	68	4	0	0	0
06-04-10	1500	100	100	99	97	95	93	90	74	10	0	0	0
06-16-10	1011	100	100	100	99	95	90	86	59	1	0	0	0
05-17-11	1600	100	72	55	44	40	37	36	33	13	1	0	0
05-26-11	0830	100	100	82	78	71	66	63	58	28	2	0	0
06-08-11	1044	100	73	65	63	61	57	54	46	5	0	0	0

Table 10. Particle-size distribution for bedload samples collected from stations in the lower Snake and Clearwater River Basins, Washington and Idaho, water years 2008–11.—Continued

[Locations of stations are shown in figure 2. **Abbreviations:** <, less than; mm, millimeter]

Date	Mean sample time	< 128 mm (percent)	< 63.0 mm (percent)	< 31.5 mm (percent)	< 16.0 mm (percent)	< 8.0 mm (percent)	< 4.0 mm (percent)	< 2.0 mm (percent)	< 1.0 mm (percent)	< 0.50 mm (percent)	< 0.25 mm (percent)	< 0.125 mm (percent)	< 0.063 mm (percent)
						South Fork Clearwater River near Harpster, Idaho (13338100)							
04-08-09	1021	100	100	100	100	100	99	95	83	34	5	0	0
04-14-09	1252	100	100	100	96	92	89	85	62	2	0	0	0
04-22-09	0937	100	100	80	78	75	70	66	44	2	0	0	0
05-07-09	0930	100	100	100	96	93	89	83	58	6	1	0	0
05-13-09	1230	100	100	100	98	87	81	76	57	13	1	0	0
05-21-09	1145	100	100	98	95	89	83	72	43	1	0	0	0
06-10-09	1414	100	100	86	82	80	75	68	39	1	0	0	0
04-22-10	1507	100	100	100	100	100	98	96	71	2	0	0	0
05-19-10	1525	100	100	100	100	99	96	89	50	2	0	0	0
05-20-10	1216	100	100	100	96	91	87	82	53	2	0	0	0
06-06-10	0840	100	100	100	99	95	87	73	37	1	0	0	0
06-14-10	1328	100	100	100	94	87	78	69	42	1	0	0	0
05-10-11	1057	100	100	100	100	100	99	96	81	53	10	0	0
06-11-11	1000	100	100	93	88	84	80	73	51	2	0	0	0
06-28-11	0903	100	100	100	100	98	94	85	49	1	0	0	0
						South Fork Clearwater River at Stites, Idaho (13338500)							
04-14-09	1646	100	100	90	90	90	90	86	64	6	1	0	0
04-22-09	1402	100	100	81	71	70	68	66	53	5	0	0	0
05-07-09	1430	100	97	71	56	45	36	33	22	1	0	0	0
05-13-09	1630	100	100	100	90	79	71	65	45	1	0	0	0
05-21-09	1430	100	100	94	80	70	61	53	35	1	0	0	0
06-10-09	0833	100	100	90	70	53	40	31	17	1	0	0	0
04-22-10	1925	100	100	100	64	56	52	48	34	1	0	0	0
05-19-10	1822	100	100	100	86	75	68	63	45	2	0	0	0
05-20-10	0847	100	100	83	79	78	76	72	57	3	0	0	0
06-05-10	1425	100	91	74	60	52	45	40	29	1	0	0	0
06-15-10	1803	100	94	94	83	69	61	53	31	1	0	0	0
05-10-11	1545	100	100	100	92	87	84	82	72	40	4	0	0
06-10-11	1555	100	74	52	39	31	24	19	13	5	0	0	0
06-28-11	1210	100	100	77	58	52	46	39	19	0	0	0	0
						Clearwater River at Orofino, Idaho (13340000)							
04-23-09	0953	100	100	100	86	78	74	72	62	5	1	0	0
05-19-09	1612	100	100	21	12	11	11	11	10	1	0	0	0
06-02-09	1809	100	94	49	23	12	10	10	9	0	0	0	0
06-04-10	0948	100	86	51	40	34	31	29	25	2	0	0	0
06-10-10	0917	100	91	52	37	31	25	22	18	0	0	0	0
06-06-11	1428	100	66	36	25	18	12	10	9	1	0	0	0
06-28-11	1518	100	100	77	74	70	65	61	57	2	0	0	0
						Clearwater River at Spalding, Idaho (13342500)							
05-05-08	1500	100	100	100	100	99	99	98	98	74	3	1	1
05-19-08	1430	100	38	33	33	33	33	33	32	26	3	0	0
05-28-08	1015	100	100	97	95	95	94	93	91	51	1	0	0
06-11-08	1345	100	100	49	31	29	28	28	27	18	0	0	0
07-07-08	1345	100	100	100	100	100	100	99	96	76	2	0	0
04-15-09	1145	100	100	100	88	87	86	85	82	1	0	0	0
04-23-09	1545	100	100	67	43	41	41	40	37	17	2	0	0
05-19-09	0955	100	100	100	86	85	85	83	76	15	2	0	0
06-03-09	0949	100	47	14	6	5	5	5	5	0	0	0	0
05-20-10	1532	100	49	29	27	27	26	26	25	2	0	0	0
06-15-11	1015	100	100	100	100	100	100	100	99	78	7	0	0

The measured bedload discharge in the Snake River near Anatone ranged from 14.2 to 491 ton/d based on 16 samples collected during 2008–11 (table 8). Based on a composite sample of all bedload sediment collected at Anatone, and using the GRADISTAT statistical software (Blott and Pye, 2001), the particle-size distribution was bimodal, with coarse gravel (16–31.5 mm) accounting for 25.7 percent of the bedload material and medium sand (0.25–0.50 mm) making up 22.0 percent (table 10). Notably absent in the bedload collected from the Snake River near Anatone was small gravel. Applying the best-fit relation between streamflow and bedload discharge (table 9) to determine the transport in the Snake River for water years 2009–11 indicates about 55,000 tons

of bedload, or about 0.62 percent of the total amount of sediment discharged from the Snake River to Lower Granite Reservoir (fig. 19).

In the Clearwater River Basin, bedload samples were collected from the Lochsa and Selway Rivers 3 and 11 times, respectively. At both stations, the particle-size distribution was bimodal, with medium sand and course gravel being the two dominant sizes (table 10). A comparison of the bedload data collected during this study from the Lochsa and Selway Rivers with data collected during 1994–97 by the U.S. Forest Service at the same sampling locations (King and others, 2004) is shown in figure 21.

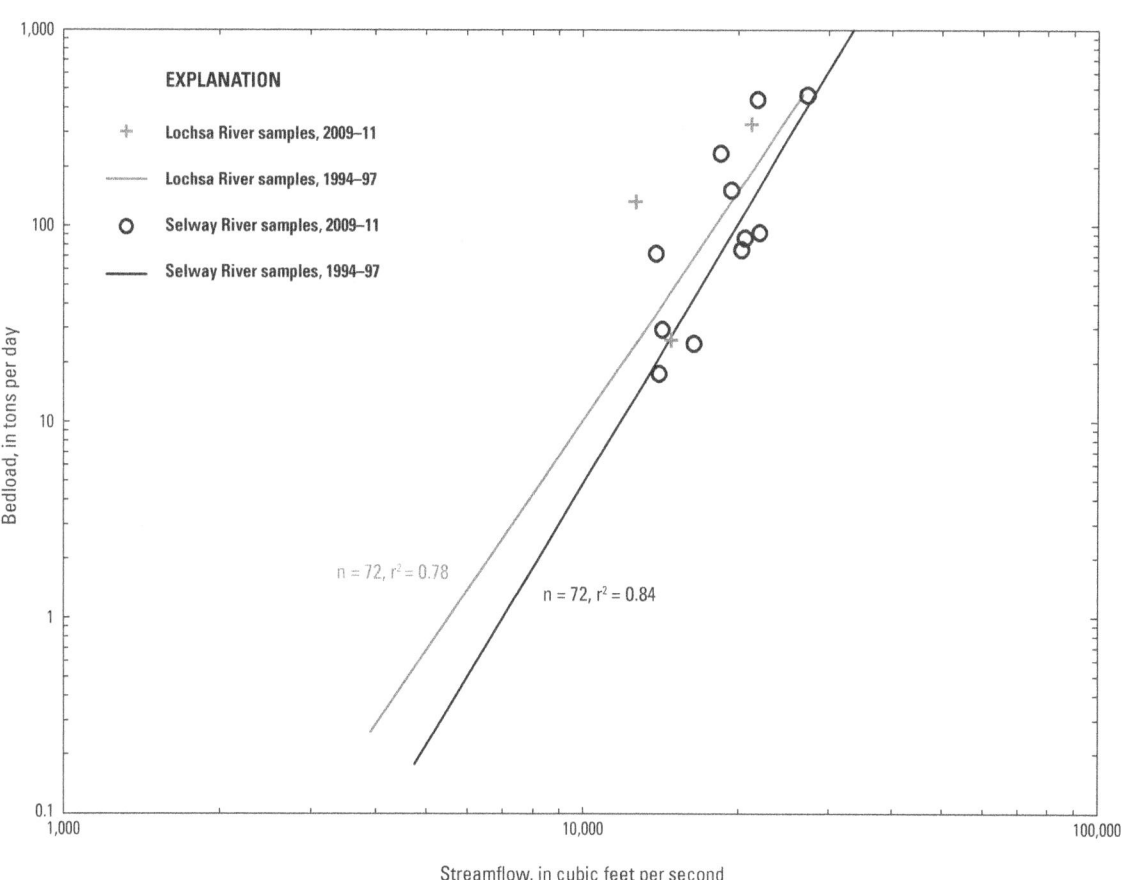

Figure 21. Bedload transport curves based on samples collected from the Lochsa and Selway Rivers near Lowell, Idaho, during water years 2009–11, and during 1994–97 by King and others (2004).

Although the range in streamflow at which bedload samples were collected was larger during the 1994–97 study, at streamflows of similar magnitude, the bedload data collected during 2009–11 corresponded well with the 1994–97 data. King and others (2004) determined that bedload discharge constituted less than 4 percent of the total sediment discharged from the Lochsa and Selway Rivers. Bedload was not estimated for the Lochsa River during 2009–11 because of the paucity of samples collected during this study. However, using the best-fit regression for the data collected from the Selway River during 2009–11, bedload constituted about 4 percent of the total sediment discharge from the Selway River (fig. 19).

Bedload samples were collected at the South Fork Clearwater River near Harpster, and South Fork Clearwater River at Stites 15 and 14 times, respectively during 2009–11 (table 10). Streamflows at Harpster ranged from 1,410 to 6,750 ft^3/s, and bedload discharge ranged from 0.6 to 103 ton/d. At the Stites station, instantaneous streamflows ranged from 2,450 to 7,490 ft^3/s and bedload ranged from 4.4 to more than 600 ton/d (table 8). Based on these data, it is apparent that although the South Fork Clearwater River at Stites is only 15 mi downstream of the Harpster station, at high streamflow the bedload transported at Stites is about one order of magnitude larger than at Harpster (fig. 20). The particle-size distribution of the bedload at both South Fork Clearwater River stations was bimodal, with the dominant size classes being medium-sized sand and medium-sized gravel. Using the equations from table 9, the bedload at Harpster constituted about 3.1 percent of the total sediment discharge during water years 2009–11, whereas at Stites the bedload constituted about 7.9 percent (figure 19).

Bedload samples were collected in the middle and lower sections of the Clearwater River at the Middle Fork Clearwater River at Kooskia, Clearwater River at Orofino, and Clearwater River at Spalding stations. Bedload discharge and bedload as a percentage of the total sediment load decreased from upstream to downstream through this reach of the Clearwater River (fig. 19) in response to a decrease in the stream gradient. The Middle Fork Clearwater River at Kooskia was sampled 11 times during 2009–11 at streamflows ranging from 22,700 to 57,000 ft^3/s (table 8). Bedload at Kooskia ranged from 13.5 to 592 ton/d, with an estimated bedload transport of 16,000 tons during water years 2009–11, approximately equivalent to the bedload discharge from the Selway River (fig. 19). The composited particle size distribution at Kooskia was bimodal, with medium sand (45 percent) and coarse gravel (12 percent) being the two dominant sizes. Bedload at the Kooskia station constituted about 3.1 percent of the total sediment load discharged from the Middle Fork Clearwater River as measured at Kooskia (fig. 19).

Bedload samples were collected seven times during 2009–11 at Clearwater River at Orofino at streamflows ranging from 37,400 to 57,700 ft^3/s. Bedload at Orofino ranged from 2.64 to 754 tons/d (table 8). The total calculated transport at Orofino was about 15,000 tons during water years 2009–11, about one-half of the combined bedload discharged from the South Fork and Middle Fork Clearwater Rivers (fig. 19). Downstream of Orofino at the Clearwater River at Spalding station, the total bedload during water years 2009–11 was about 9,500 tons, less than 1 percent of the total sediment transport at the station during that period. The reach of the Clearwater River from Orofino to Spalding probably transports less sediment (both suspended load and bedload) than it did historically because of the construction of Dworshak Reservoir, which essentially negates sediment delivery to the main stem Clearwater River from the North Fork Clearwater River. The total bedload for 2009–11 in the Clearwater River at Spalding was about 9,500 tons, about 15 percent of the total bedload discharged to Lower Granite Reservoir from the combined Clearwater and Snake Rivers. Overall, bedload accounted for only about 0.64 percent of the total sediment load entering Lower Granite Reservoir during water years 2009–11 (fig. 19).

Bedload data collected from the Snake River near Anatone and the Clearwater River at Spalding during 1972–79 (Jones and Seitz, 1980) indicated that historically, bedload was larger and comprised a larger part of the overall sediment load. Based on 63 samples collected during 1972–79 from the Snake River near Anatone at streamflows ranging from 27,500 to 161,000 ft^3/s, the mean bedload was about 450 ton/d and ranged from less than 1.0 to about 5,600 ton/d. In comparison, the mean bedload based on 16 samples collected at Anatone during this study was about 130 ton/d and ranged from 14.2 to 491 ton/d (table 8). Although the magnitude of streamflows sampled during both study periods were similar, the mean bedload was about 3.5 times higher during 1972–79. Jones and Seitz (1980) reported that bedload comprised about 4 percent of the overall sediment load transported in the Snake River during 1972–79, whereas results from this study indicate that bedload comprised less than 1 percent of the sediment load. Although the relation between streamflow and bedload was relatively poor for the 1972–79 (R^2=0.44) and 2008–11 data (R^2=0.26), there is an apparent shift in the bedload transport curve, indicating less bedload in the Snake River near Anatone during 2008–11 (fig. 22). The reason for this apparent shift is difficult to determine given the increase in the transport of suspended sand during 2008–11. Additional bedload data collected during 2008-11 from the Snake River near Anatone would be helpful to verify the apparent decrease in bedload in the Snake River.

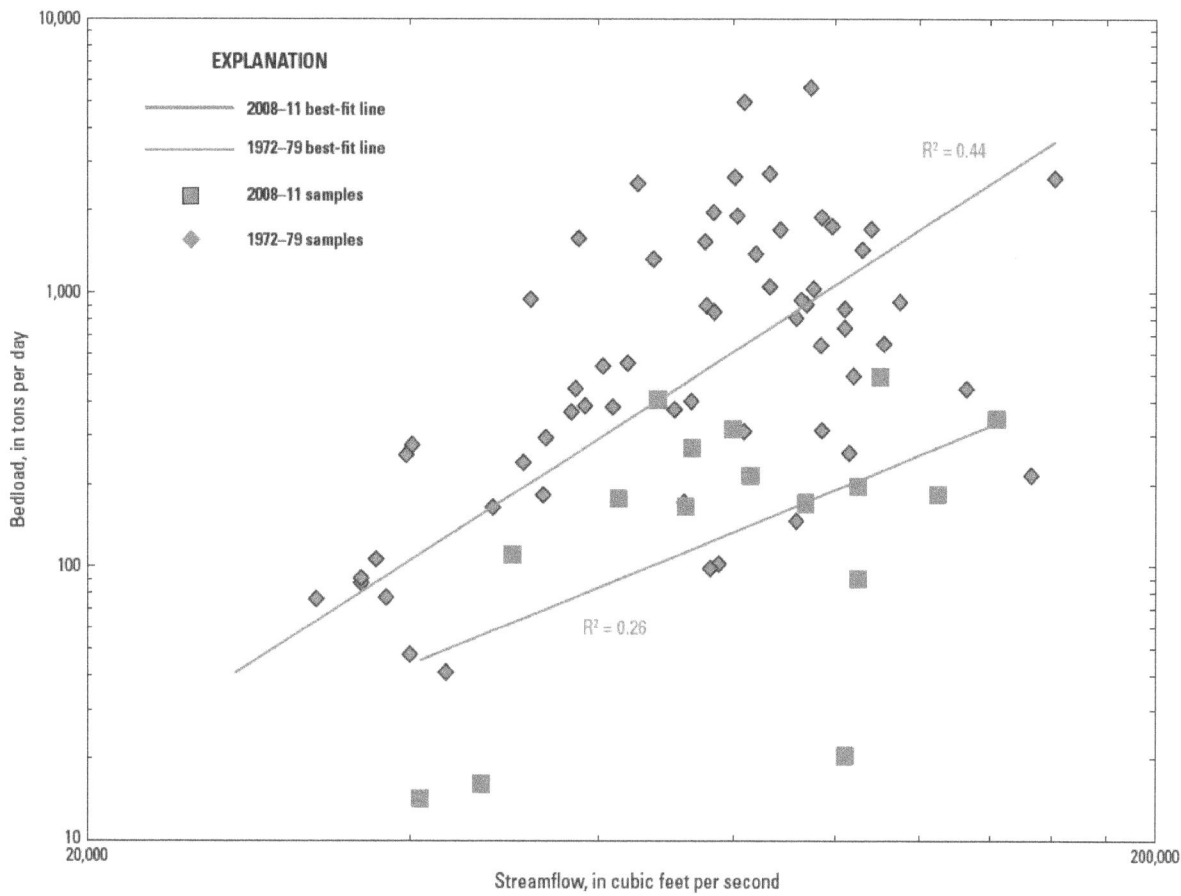

Figure 22.   Bedload transport curves comparing data collected at Snake River near Anatone, Washington, during water years 1972–79 and 2008–11.

Similarly, a comparison of data from 1972–79 and 2008–11 indicated a decrease in bedload in the Clearwater River at Spalding (fig. 23). The mean bedload during 1972–79 based on 78 samples collected at the Spalding station was about 120 ton/d, ranging from less than 1.0 to about 3,700 ton/d. The mean bedload based on 11 samples collected during 2008–11 was about 54 ton/d, ranging from about 6.0 to 840 ton/d (table 8). Based on this comparison, bedload in the Clearwater River during 1972–79 was markedly larger (mean

of about 2.2 times) at roughly equivalent stream discharge. Jones and Seitz (1980) reported that, similar to the Snake River, during 1972–79 bedload comprised about 4 percent of the total sediment load in the Clearwater River. The 2008–11 data indicate that bedload was less than 1 percent of the sediment load in the Clearwater River at Spalding. The relation between streamflow and bedload was better (higher R^2) for the Clearwater River than the Snake River for both sampling periods (figs. 22 and 23).

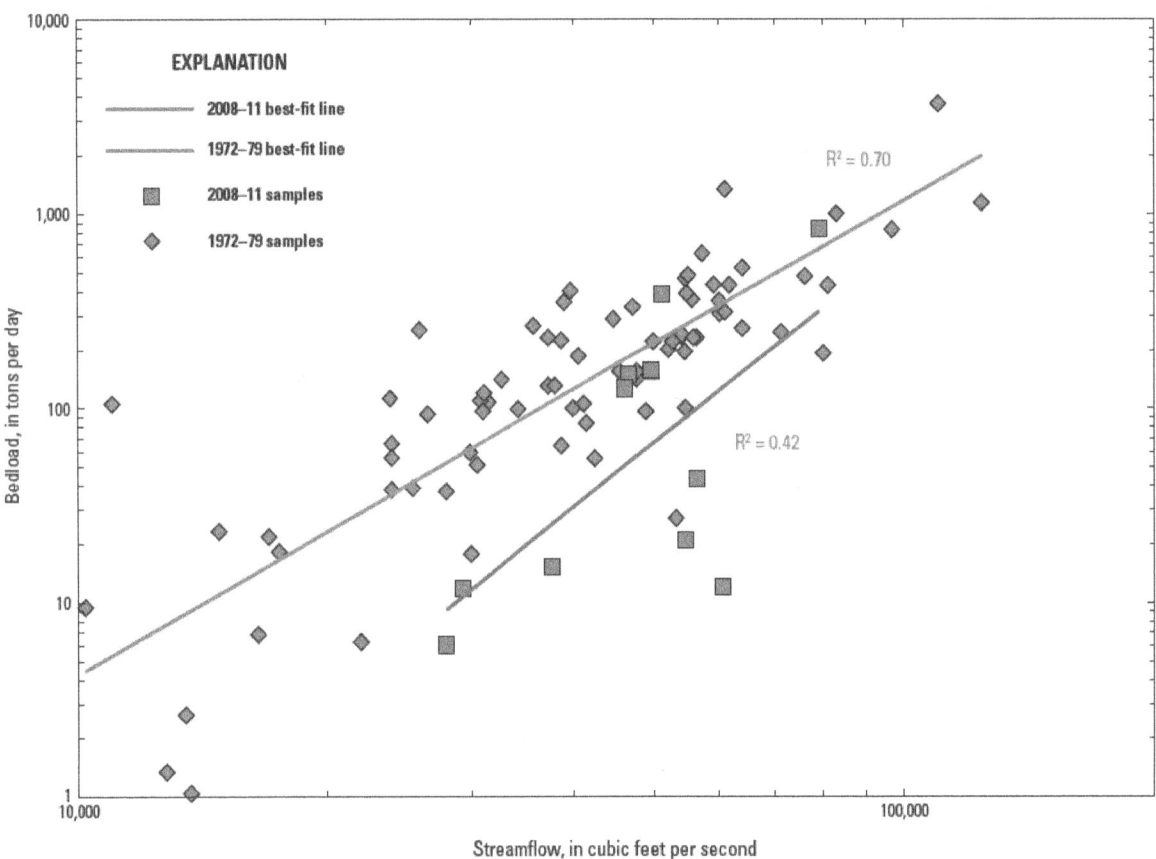

Figure 23.   Bedload transport curves comparing data collected at Clearwater River at Spalding, Idaho, water years 1972–79 and 2008–11.

Summary

Since Lower Granite Dam was completed in 1975, about 75 million cubic yards of sediment have accumulated in Lower Granite Reservoir, an average annual accumulation of about 2.3 million cubic yards through 2008. In 2008, the U.S. Geological Survey (USGS), in cooperation with the U.S. Army Corps of Engineers (USACE), conducted sediment sampling in the Snake River near Anatone, Washington, and the Clearwater River at Spalding, Idaho, to quantify sediment loading to Lower Granite Reservoir and to evaluate sediment depositional characteristics in the reservoir. In 2009, sediment sampling was extended to 10 additional sampling stations in the lower Snake and Clearwater River basins to help identify the subbasins contributing most of the sediment being delivered to Lower Granite Reservoir.

Of the stations sampled during 2009–11, the largest measured total suspended sediment (TSS) concentrations were in samples collected from the Potlatch (3,300 mg/L) and Palouse (1,400 mg/L) Rivers during a rain-on-snow event that occurred in January 2011. The largest median concentration of TSS (100 mg/L) and the largest suspended sediment fraction as fine-grained silt and clay (median of 95 percent) were in samples collected from the Palouse River. Other rivers with a large percentage of fine-grained suspended sediment were the Potlatch (median of 92 percent) and the Grand Ronde (median of 84 percent) Rivers. The Palouse, Potlatch, and Grande Ronde Rivers all drain basins with relatively large proportions of agricultural activity. The Selway and Lochsa Rivers, the Middle Fork Clearwater River at Kooskia, and the Clearwater River at Orofino generally had small concentrations of TSS with medians of 11, 11, 15, and 13 mg/L, respectively.

During water years 2009–11, Lower Granite Reservoir received about 10 million tons of suspended sediment from the combined loads of the Snake and Clearwater Rivers. About 60 percent of sediment entering the reservoir during 2009–11 occurred during water year 2011, which was characterized by a large winter snowpack followed by sustained spring runoff and high suspended-sediment transport during most of the summer. Of the sediment load entering the reservoir, the Snake River accounted for about 89 percent of the TSS, about 90 percent of the suspended sand, and about 87 percent of the suspended fines. The Salmon River contributed about 51 percent of the TSS, about 56 percent of the suspended sand, and about 44 percent of the suspended fines transported to Lower Granite Reservoir.

A comparison of historical sediment data collected from the Snake River with data collected during this study indicated that concentrations of TSS and suspended sand were significantly larger during 2008–11 compared to 1972–79, whereas the concentrations of suspended fines were not. In the Snake River, the sand fraction increased from an average of 28 percent of the TSS load during 1972–79 to an average of 48 percent during 2009–11. The increase in the suspended-sand load in the Snake River is probably attributable to numerous severe forest fires that burned large areas of central Idaho from 1980–2010. Additional fluvial-sediment monitoring in the Salmon River Basin would be helpful to identify source areas and to quantify the magnitude of sediment delivery to discrete reaches of the Salmon River, the lower Snake River, and Lower Granite Reservoir.

In the Clearwater River, data collected during this study indicated that the TSS and suspended fines concentrations during 1972–79 were not significantly different from the concentrations during 2008–11. However, the concentrations of suspended sand in the Clearwater River were significantly larger during 2008–11. The increase in the sand load in the Clearwater River may also be attributable to forest fire activity in areas of the basin with highly erodible soils.

Suspended-sediment surrogate models developed using acoustic Doppler velocity meters (ADVMs) effectively estimated suspended-sediment concentrations (SSCs) and loads for most streamflow conditions in the lower Snake and Clearwater Rivers. At both stations (lower Snake River near Anatone and Clearwater River at Spalding) instrumented with ADVMs, surrogate models developed using acoustic backscatter had a better correlation than LOADEST models for SSC and suspended fines. Surrogate models also had a better correlation for suspended sands at the Spalding station

on the lower Clearwater River. Over short (monthly and storm event) and long (annual) time scales when sediment concentrations and loads are highly variable, acoustic backscatter appears to provide better estimates of sediment concentration and load than do traditional sediment transport curves based solely on streamflow. Because acoustic backscatter is not affected by hysteresis, it provides a direct, in-situ measurement of suspended sediment, and is effective in representing sediment sources from a combination of regulated and unregulated sources. On a monthly and (or) annual basis, the acoustic surrogate and LOADEST models appear to produce similar sediment load estimates for the Clearwater River at Spalding. However, estimates of SSCs and loads in the Snake River differed substantially between the acoustic surrogate and LOADEST models. The largest difference occurred in water year 2011, when the suspended-sediment load calculated using the LOADEST model was more than three times higher than the suspended-sediment load calculated using the acoustic surrogate model. The largest discrepancies between the LOADEST and acoustic surrogate models occur primarily during high streamflow, when the SSC and the suspended sand fraction are large. This discrepancy may be a result of acoustic surrogate tools being unable to capture the contribution of suspended sand moving near the bottom of the water column.

Bedload accounted for less than 1 percent of the total sediment load entering Lower Granite Reservoir from the Snake and Clearwater Rivers. The estimated bedload in the Salmon River at White Bird during 2009–11 was about 66,000 tons, the largest amount of the stations sampled. The lower Snake River basin, which includes the Salmon River, had the second largest measured bedload with a total of 55,000 tons during 2009–11, about 0.62 percent of the total sediment load entering Lower Granite Reservoir from the Snake River. The estimated bedload in the Clearwater River at Spalding was only about 9,500 tons, roughly 0.83 percent of the total sediment load transported to Lower Granite Reservoir from the Clearwater River.

Acknowledgments

The USGS wishes to thank the U.S. Army Corps of Engineers, Walla Walla District for providing funding for this project. Special thanks go to USGS field personnel from the Idaho Water Science Center Boise and Post Falls Field Offices who were responsible for most of the data collection.

References Cited

Aikaike, Hirotugu, 1981, Likelihood of a model and information criteria: Journal of Econometrics, v. 16, no. 1, p. 3–14.

Blott, S.J., and Pye, Kenneth, 2001, GRADISTAT—A grain size distribution and statistics package for the analysis of unconsolidated sediments: Earth Surface Processes and Landforms, v. 26, p. 1237–1248.

Bradu, Dan, and Mundlak, Yair, 1970, Estimation in lognormal linear models: Journal of the American Statistical Association, v. 65, no. 329, p. 198–211.

Braun, C.L., Wilson, J.T., Van Metre, P.C., Weakland, R.J., Fosness, R.L., and Williams, M.L, 2012, Grain-size distribution and selected major and trace element concentrations in bed-sediment cores from Lower Granite Reservoir and Snake and Clearwater Rivers, eastern Washington and northern Idaho, 2010: U.S. Geological Survey Scientific Investigations Report 2012-5219, 81 p.

Burton, T.A., 2005, Fish and stream habitat risks from uncharacteristic wildfire—Observations from 17 years of fire-related disturbances on the Boise National Forest, Idaho: Forest Ecology and Management, v. 211, p. 140–149.

Chanson, Hubert, Takeuchi, Maiko, and Trevethan, Mark, 2008, Using turbidity and acoustic backscatter intensity as surrogate measures of suspended sediment concentration in a small subtropical estuary: Journal of Environmental Management, v. 88, p. 1406–1416.

Cohn, T.A., Caulder, D.L., Gilroy, E.J., Zynjuk, L.D., and Summers, R.M., 1992, The validity of a simple statistical model for estimating fluvial constituent loads—An empirical study involving nutrient loads entering Chesapeake Bay: Water Resources Research, v. 28, no. 9, p. 2353–2363.

Cohn, T.A., Delong, L.L, Gilroy, E.J., Hirsch, R.M., and Wells, D.K., 1989, Estimating constituent loads: Water Resources Research, v. 25, no. 5, p. 937–942.

Crawford, C.G., 1991, Estimation of suspended-sediment rating curves and mean suspended-sediment loads: Journal of Hydrology, v. 129, p. 331–348.

Davis, B.E., 2005, A guide to the proper selection and use of federally approved sediment and water-quality samplers: U.S. Geological Survey Open-File Report 2005-1087, 20 p. (Also available at http://pubs.er.usgs.gov/publication/ofr20051087.)

Duan, N.,1983, Smearing estimate: A nonparametric retransformation method: Journal of the American Statistical Association, v.78, no. 383, p. 605–610.

Ebbert, J.C., and Roe, R.D., 1998, Soil erosion in the Palouse River Basin—Indications of improvement: U.S. Geological Survey Fact Sheet 069-98. (Also available at http://wa.water.usgs.gov/pubs/fs/fs069-98/.)

Edwards, T.K., and Glysson, G.D., 1999, Field methods for measurement of fluvial sediment: U.S. Geological Survey Techniques of Water-Resources Investigations, book 3, chap. C2, 89 p. (Also available at http://pubs.usgs.gov/twri/twri3-c2/.)

Elci, S., Aydln, R., and Work, P.A., 2009, Estimation of suspended sediment concentration in rivers using acoustic methods: Environmental Monitoring and Assessment, v 59, p. 255–265.

Emmett, W.W., 1976, Bedload transport in two large, gravel-bed rivers, Idaho and Washington: Proceedings, Third Federal Inter-Agency Sedimentation Conference, Denver, Colo., p. 4-101–4-114.

Folk, R.L., and Ward, W.C., 1957, Brazos River bar—A study in the significance of grain size parameters: Journal of Sedimentary Petrology, v. 27, p. 3–26.

Gartner, J.W., 2004, Estimating suspended solids concentrations from backscatter intensity measured by acoustic Doppler current profiler in San Francisco Bay, California: Marine Geology, v. 211, p. 169–187.

Gartner, J.W., and Gray, J.R., 2005, Summary of suspended-sediment technologies considered at the Interagency Workshop on Turbidity and Other Sediment Surrogates, in Gray, J.R., ed., Proceedings of the Federal Interagency Sediment Monitoring Instrument and Analysis Research Workshop, September 9–11, 2003, Flagstaff, Arizona: U.S. Geological Survey Circular 1276, 9 p. (Also available at http://pubs.er.usgs.gov/publication/cir1276.)

Gilroy, E.J., Hirsch, R.M., and Cohn, T.A., 1990, Mean square error of regression-based constituent transport estimates: Water Resources Research, v. 26, p. 2069–2088.

Goolsby, D.A., Battaglin, W.A., Lawrence, G.B., Artz, R.S., Aulenbach, B.T., Hooper, R.P., Keeney, D.R., and Stensland, G.J., 1999, Flux and sources of nutrients in the Mississippi-Atchafalaya River Basin—Topic 3 Report for the Integrated Assessment on Hypoxia in the Gulf of Mexico: Silver Spring, Md., National Oceanic and Atmospheric Administration Coastal Ocean Program Decision Analysis Series No. 17, 130 p.

Guy, H.P., 1969, Laboratory theory and methods for sediment analysis: U.S. Geological Survey Techniques of Water-Resources Investigations, book 5, chap. C1, 58 p. (Also available at http://pubs.usgs.gov/twri/twri5c1/.)

Helley, E.J., and Smith, Winchell, 1971, Development and calibration of a pressure-difference bedload sampler: U.S. Geological Survey Open-File Report, 18 p.

Helsel, D.R., and Hirsch, R.M., 1992, Statistical methods in water resources: New York, Elsevier, 522 p.

Hubbell, D.W., 1964, Apparatus and techniques for measuring bedload: U.S. Geological Survey Water-Supply Paper 1748, 74 p. (Also available online at http://pubs.er.usgs.gov/publication/wsp1748.)

Jones, M.L., and Seitz, H.R., 1980, Sediment transport in the Snake and Clearwater Rivers in the vicinity of Lewiston, Idaho: U.S. Geological Survey Open-File Report 80-690, 179 p.

Judge, C.G., Griffiths, W.E., Hill, R.C., Lutkepohl, H., and Lee, T.C., 1985, The theory and practice of econometrics: New York, John Wiley and Sons, p. 870–873.

King, J.G., Emmett, W.W., Whiting, P.J., Kenworthy, R.P., and Barry, J.J., 2004, Sediment transport data and related information for selected coarse-bed streams and rivers in Idaho: Fort Collins, Colo., U.S. Department of Agriculture, Forest Service, Rocky Mountain Research Station, Gen. Tech. Rep. RMRS-GTR-131, 26 p.

Latah Soil and Water Conservation District, 2007, Potlatch River Watershed Management Plan: Latah Soil and Water Conservation District Web site, accessed February 7, 2013, at http://latahsoil.org/id50.html.

Miller, S., Glanzman, R., Doran, S., Parkinson, S.K., Buffington, J., and Milligan, J., 2003, Geomorphology of the Hells Canyon Reach of the Snake River, in Technical appendices for Hells Canyon Hydroelectric Project: Boise, Id., Idaho Power Technical Report E.1-2.

Mueller, D.S., and Wagner, C.R., 2009, Measuring discharge with acoustic Doppler current profilers from a moving boat: U.S. Geological Survey Techniques and Methods, book 3, chap. A22, 72 p.

Ott, Lyman, and Longnecker, Michael, 2001, An introduction to statistical methods and data analysis, (5th ed.): Pacific Grove, California, Duxbury, 1152 p.

Parkinson, S., Anderson, K., Conner, J., and Milligan J., 2003, Sediment transport, supply and stability in the Hells Canyon reach of the Snake River: Idaho Power Technical Report for FERC No. 1971, 140 p.

Patino, E., and Byrne, M.J., 2004, Application of acoustic and optic methods for estimating suspended-solids concentrations in the St. Lucie River Estuary, Florida: U.S. Geological Survey Scientific Investigations Report 2004-5028, 23 p. (Also available at http://pubs.usgs.gov/sir/2004/5028/.)

Runkel, R.L., Crawford, C.G., and Cohn, T.A., 2004, Load Estimator (LOADEST)—A FORTRAN program for estimating constituent loads in streams and rivers: U.S. Geological Survey Techniques and Methods, book 4, chap. A5, 69 p. (Also available at http://pubs.er.usgs.gov/publication/tm4A5.)

Schumacher, B.A., 2002, Methods for the determination of total organic carbon (TOC) in soils and sediments: U.S. Environmental Protection Agency publication NCEA-C-1282, 23 p.

Smith, R.A., Alexander, R.B., and Wolman, M.G., 1987, Water-quality trends in the Nation's rivers: Science, v. 235, p. 1607–1615.

Teasdale, G.N., 2010, Sediment load, transport and accumulation in Lower Granite Reservoir on the Snake River: 2nd Joint Federal Interagency Conference, Las Vegas, Nevada, June 27–July 1, 2010.

Tetra Tech, 2006, Investigation of sediment source and yield, management, and restoration opportunities within the lower Snake River Basin: Submitted to Walla Walla District, U.S. Army Corps of Engineers, Delivery Order No. 7, Contract W912EF-05-D-002.

Topping, D., Melis, T., Rubin, D., and Wright, S.A., 2004, High-resolution monitoring of suspended-sediment concentration and grain size in the Colorado River using laser-diffraction instruments and a three-frequency acoustic system, in Cheng Liu, ed., Proceedings of the 9th International Symposium on River Sedimentation, October 18–21, 2004: Yichang, China, International Sediment Initiative, p. 2507–2514.

Turnipseed, D.P., and Sauer, V.B., 2010, Discharge measurements at gaging stations: U.S. Geological Survey Techniques and Methods book 3, chap. A8, 87 p. (Also available at http://pubs.er.usgs.gov/publication/tm3A7.)

U.S. Army Corps of Engineers, 2002, Dredged Material Management Plan and Environmental Impact Statement, McNary Reservoir and Lower Snake River Reservoirs, appendix A—Hydrologic Analysis: U.S. Army Corps of Engineers, 27 p.

U.S. Army Corps of Engineers, 2005, Intent to prepare a draft Environmental Impact Statement, Programmatic Sediment Management Plan, lower Snake River reservoirs, in the states of Washington and Idaho: Federal Register, v. 70, no. 190, p. 57569–57570.

U.S. Geological Survey, 2013, USGS surface-water data for Idaho: U.S. Geological Survey database, accessed February 7, 2013, at http://waterdata.usgs.gov/id/nwis/sw.

Wall, G.R., Nystrom, E.A., and Litten, Simon, 2006, Use of an ADCP to compute suspended-sediment discharge in the tidal Hudson River, New York: U.S. Geological Survey Scientific Investigations Report 2006-5055, 16 p. (Also available at http://pubs.er.usgs.gov/publication/sir20065055.)

Washington State Department of Ecology, 2006, Palouse River Watershed: Washington State Department of Ecology Focus Sheet 06-10-002, 2 p.

Williams, M.L., Fosness, R.L., and Weakland, R.J., 2012, Bathymetric and underwater video survey of Lower Granite Reservoir and vicinity, Washington and Idaho, 2009–10: U.S. Geological Survey Scientific Investigations Report 2012-5089, 22 p. (Also available at http://pubs.er.usgs.gov/publication/sir20125089.)

Wood, M.S., and Teasdale, G.N., 2013, Use of surrogate technologies to estimate suspended sediment in the Clearwater River, Idaho, and Snake River, Washington, 2008–10: U.S. Geological Survey Scientific Investigations Report 2013-5052, 30 p. (Also available at http://pubs.usgs.gov/sir/2013/5052/.)